Time–Frequency Analysis of Seismic Signals

Time–Frequency Analysis of Seismic Signals

Yanghua Wang
Imperial College London

WILEY

This edition first published 2023

Registered Offices
John Wiley & Sons, Inc., 111 River Street, Hoboken, NJ 07030, USA
John Wiley & Sons Ltd, The Atrium, Southern Gate, Chichester, West Sussex, PO19 8SQ, UK

Editorial Office
The Atrium, Southern Gate, Chichester PO19 8SQ, UK

For details of our global editorial offices, customer services, and more information about Wiley products visit us at www.wiley.com.

Wiley also publishes its books in a variety of electronic formats and by print-on-demand. Some content that appears in standard print versions of this book may not be available in other formats.

Library of Congress Cataloging-in-Publication Data applied for

[HB ISBN: 9781119892342]

Cover Design: Wiley
Cover Image: Courtesy of Yanghua Wang

Set in 10/14pt MeridienLTStd by Straive, Chennai, India

Printed and bound by CPI Group (UK) Ltd, Croydon, CR0 4YY

C9781119892342_150922

This book is dedicated to my wife, Guo-ling, and my two children, Brian and Claire.

Contents

Preface ix

1 Nonstationary Signals and Spectral Properties 1

1.1 Stationary Signals 1
1.2 Nonstationary Signals 5
1.3 The Fourier Transform and the Average Properties 7
1.4 The Analytic Signal and the Instantaneous Properties 10
1.5 Computation of the Instantaneous Frequency 13
1.6 Two Groups of Time–Frequency Analysis Methods 16

2 The Gabor Transform 19

2.1 Short-time Fourier Transform 19
2.2 The Gabor Transform 23
2.3 The Cosine Function Windows 26
2.4 Spectral Leakage of Window Functions 31
2.5 The Gabor Limit of Time–Frequency Resolution 33
2.6 Implementation of the Gabor Transform 36
2.7 The Inverse Gabor Transform 40
2.8 Application in Inverse Q Filtering 42

3 The Continuous Wavelet Transform 47

3.1 Basics of the Continuous Wavelet Transform 47
3.2 The Complex Morlet Wavelet 51
3.3 The Complex Morse Wavelet 54
3.4 The Generalised Seismic Wavelet 58
3.5 The Pseudo-frequency Representation 62
3.6 The Inverse Wavelet Transform 65
3.7 Implementation of the Continuous Wavelet Transform 67
3.8 Hydrocarbon Reservoir Characterisation 69

4 The S Transform 73

4.1 Basics of the S Transform 74
4.2 The Generalised S Transform 77
4.3 The Fractional Fourier Transform 79
4.4 The Fractional S Transform 83
4.5 Implementation of the S Transforms 86
4.6 The Inverse S Transforms 88
4.7 Application to Clastic and Carbonate Reservoirs 93

5 The W Transform 95

5.1 Basics of the W Transform 95
5.2 The Generalised W Transform 99

5.3 Implementation of Nonstationary Convolution 106
5.4 The Inverse W Transform 108
5.5 Application to Detecting Hydrocarbon Reservoirs 109
5.6 Application to Detecting Karst Voids 112

6 The Wigner–Ville Distribution 117

6.1 Basics of the Wigner–Ville Distribution (WVD) 117
6.2 Defining the WVD with an Analytic Signal 120
6.3 Properties of the WVD 123
6.4 The Smoothed WVD 126
6.5 The Generalised Class of Time–Frequency Representations 132
6.6 The Ambiguity Function and the Generalised WVD 134
6.7 Implementation of the Standard and Smoothed WVDs 140
6.8 Implementation of the Ambiguity Function and the Generalised WVD 147

7 Matching Pursuit 151

7.1 Basics of Matching Pursuit 151
7.2 Three-stage Matching Pursuit 153
7.3 Matching Pursuit with the Morlet Wavelet 157
7.4 The Sigma Filter 159
7.5 Multichannel Matching Pursuit 163
7.6 Structure-adaptive Matching Pursuit 168
7.7 Three Applications 170

8 Local Power Spectra with Multiple Windows 175

8.1 Multiple Orthogonal Windows 176
8.2 Multiple Windows Defined by the Prolate Spheroidal Wavefunctions 178
8.3 Multiple Windows Constructed by Solving a Discretised Eigenvalue Problem 180
8.4 Multiple Windows Constructed by Gaussian Functions 184
8.5 The Gabor Transform with Multiple Windows 187
8.6 The WVD with Multiple Windows 191
8.7 Prospective of Time–Frequency Analysis without Windowing 195

Appendices 197

A The Gaussian Integrals, the Gamma Function, and the Gaussian Error Functions 197
B Fourier Transforms of the Tapered Boxcar Window, the Truncated Gaussian Window, and the Weighted Cosine Window 200
C The Generalised Seismic Wavelet in the Time Domain 203
D Implementation of the Fractional Fourier Transform 205
E Marginal Properties and the Analytic Signal in the WVD Definition 206
F The Prolate Spheroidal Wavefunctions, the Associated and the Ordinary Legendre Polynomials 211

References 219

Author Index 227

Subject Index 229

Preface

The aim of this book, *Time–Frequency Analysis of Seismic Signals*, is to reveal the local properties of nonstationary seismic signals. Their local properties, such as time period, frequency, and spectral content, vary with time, and the time of seismic signals is a proxy for geologic depth.

Time–frequency analysis is a generalisation of classical Fourier spectral analysis for stationary signals, by extending it to spectral analysis of nonstationary signals. All methods of time–frequency analysis can be treated as a series of implementations of the classical Fourier transform. The kernel of each Fourier transform is formed from a segment of the original seismic signal. Therefore, the time–frequency spectrum is composed of the Fourier spectra generated at different time positions, where the time position corresponds to the centre of each segment.

Different methods of time–frequency analysis differ in the construction of the local kernel, before the Fourier transform is applied. Because of the differences in the construction of the Fourier transform kernel, the methods of time–frequency analysis in this book are divided into two groups.

The first group of methods are the Gabor transform-type methods, including the Gabor transform itself (Chapter 2), the continuous wavelet transform (Chapter 3), the S transform (Chapter 4), and the W transform (Chapter 5). This group of methods focuses on the properties of the different forms of window functions. The window function, which acts as a convolution operator, is applied to the nonstationary signal in a sliding manner. The windowed signal segment is the local kernel for the Fourier transform.

The second group of methods are the energy density distribution methods, including the Wigner–Ville distribution (WVD) (Chapter 6), the matching pursuit-based WVD (Chapter 7), and the local power spectrum with multiple orthogonal windows (Chapter 8). While the resulting time–frequency representation is the energy density distribution, this second group of methods focuses on manipulating and reformulating the original signal segment to form a time-varying kernel for the Fourier transform.

An essential difference between these two groups of methods is that the Gabor transform-type methods produce complex spectra in the time–frequency plane, while the energy density distribution methods produce real spectra that do not contain phase information. The Gabor transform-type methods have an inverse transform counterpart to reconstruct the original nonstationary signal, and therefore the time–frequency spectra produced by the Gabor transform-type methods can be used in signal processing. However, the energy density distribution methods cannot recover the original signal because the phase information in the time–frequency plane is missing. Consequently, the time–frequency representation of the energy density distribution methods can only be used as a seismic attribute for geophysical characterisation.

With respect to the collective groupings, this book systematically presents various time–frequency analysis methods, including some techniques that have not yet been published or that the author has been instrumental in developing. In presenting each method, the basic theory and mathematical concepts are summarised, with emphasis on the technical aspects.

As such, this book is a practical guide for geophysicists seeking to produce geophysically meaningful time–frequency spectra, process seismic data with time-dependent operations to faithfully represent nonstationary signals, and use the seismic attributes of time–frequency space for the quantitative characterisation of hydrocarbon reservoirs.

Yanghua Wang

25 July 2022

1

Nonstationary Signals and Spectral Properties

Stationary signals are any idealised signals that have time-independent properties, such as time period, frequency, and spectral content. Seismic signals, however, are nonstationary signals that violate the stationary rule described above. When seismic waves propagate through the anelastic media in the Earth's subsurface, seismic signals have variable properties that vary with the propagation path and travel time.

A stationary signal can be represented mathematically by a stack of sinusoids, using the Fourier transform. In contrast, a nonstationary signal cannot be properly represented by the Fourier transform because the amplitude and frequency of a sinusoidal representation can change dynamically as a function of travel time. To investigate the local properties of a nonstationary signal, an analytic signal-based analysis method can be used instead.

1.1 Stationary Signals

For seismic signals, an idealised stationary model is a stationary convolution process. The physical meaning of the convolution process is 'superposition', which is a fundamental principle in the analysis of seismic signals.

Consider an earth model with a layered structure (Figure 1.1). Each layer has a different acoustic impedance, which is the product of velocity and density. The contrast of this physical property between two layers causes seismic reflections at the interface.

Time–Frequency Analysis of Seismic Signals, First Edition. Yanghua Wang.

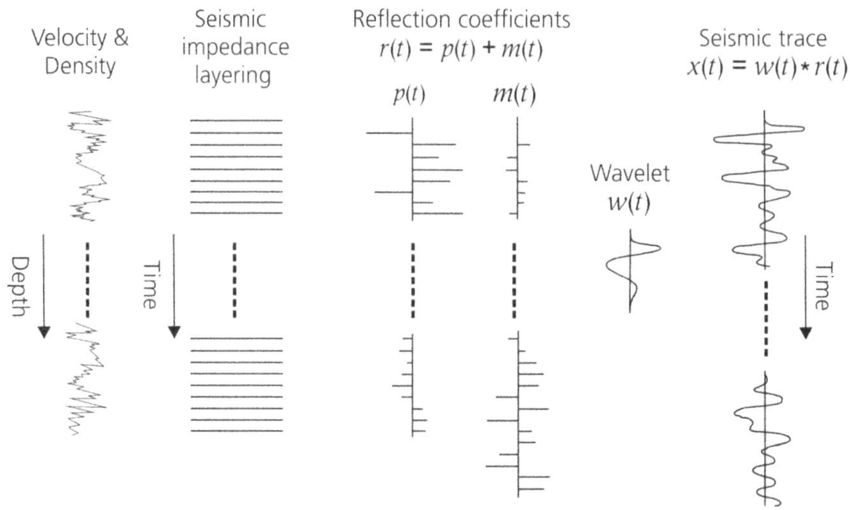

Figure 1.1 The principle of 'superposition'. Each wavelet is scaled by the reflectivity, which is the contrast in the acoustic impedance between two layers. A seismic trace $x(t)$ is formed by summing all scaled wavelets, which are reflected from various interfaces.

The contrast of acoustic impedance is called reflectivity, or reflection coefficient. The seismic reflectivity includes primary $p(t)$ and multiple $m(t)$, where t is the travel time. The reflectivity series in time consists of both types, $r(t) = p(t) + m(t)$. Each reflectivity serves as a scaling factor to scale a particular wavelet $w(t)$. A recorded seismic trace $x(t)$ is formed by summing all scaled wavelets reflected from different interfaces.

This physical process is the 'superposition' and can be written as

$$\begin{bmatrix} x_0 \\ x_1 \\ \vdots \\ \vdots \\ \vdots \\ \vdots \\ x_{\ell-1} \\ \vdots \\ x_{\ell+n-2} \end{bmatrix} = r_0 \begin{bmatrix} w_0 \\ w_1 \\ \vdots \\ w_{n-1} \\ \\ \\ \\ \\ \\ \end{bmatrix} + r_1 \begin{bmatrix} \\ w_0 \\ w_1 \\ \vdots \\ w_{n-1} \\ \\ \\ \\ \\ \end{bmatrix} + r_2 \begin{bmatrix} \\ \\ w_0 \\ w_1 \\ \vdots \\ w_{n-1} \\ \\ \\ \\ \end{bmatrix} + \cdots + r_{\ell-2} \begin{bmatrix} \\ \\ \\ \\ w_0 \\ w_1 \\ \vdots \\ w_{n-1} \\ \\ \end{bmatrix} + r_{\ell-1} \begin{bmatrix} \\ \\ \\ \\ \\ w_0 \\ w_1 \\ \vdots \\ w_{n-1} \end{bmatrix}, \tag{1.1}$$

where $\mathbf{r} = [r_0, r_1, r_2, \cdots, r_{\ell-1}]^{\mathrm{T}}$ is the vector of the reflectivity series, $\mathbf{w} = [w_0, w_1, w_2, \cdots, w_{n-1}]^{\mathrm{T}}$ is the vector of the seismic wavelet, and $\mathbf{x} = [x_0, x_1, x_2, \cdots, x_{\ell+n-2}]^{\mathrm{T}}$ is the vector of the seismic trace. If the length of the reflectivity series is ℓ, and the length of the seismic wavelet is n, the length of the resulting seismic trace is $\ell + n - 1$.

In the superposition process of Eq. (1.1), the wavelet is time shifted and scaled by the reflectivity. Combining all the time-shifted wavelets to form a wavelet matrix, $\mathbf{W} = [\mathbf{w}_0, \mathbf{w}_1, \mathbf{w}_2, \cdots, \mathbf{w}_{\ell-1}]$, the superposition process can be expressed in a matrix–vector form, $\mathbf{x} = \mathbf{Wr}$, and written explicitly as

$$\begin{bmatrix} x_0 \\ x_1 \\ x_2 \\ \vdots \\ \vdots \\ x_{\ell-2} \\ x_{\ell-1} \\ x_\ell \\ \vdots \\ x_{\ell+n-3} \\ x_{\ell+n-2} \end{bmatrix} = \begin{bmatrix} w_0 & & & & & & \\ w_1 & w_0 & & & & & \\ w_2 & w_1 & w_0 & & & & \\ \vdots & w_2 & w_1 & \ddots & & & \\ w_{n-1} & \vdots & w_2 & \ddots & \ddots & & \\ & w_{n-1} & \vdots & \ddots & \ddots & w_0 & \\ & & w_{n-1} & & \ddots & w_1 & w_0 \\ & & & \ddots & & w_2 & w_1 \\ & & & & \ddots & \vdots & w_2 \\ & & & & & w_{n-1} & \vdots \\ & & & & & & w_{n-1} \end{bmatrix} \begin{bmatrix} r_0 \\ r_1 \\ r_2 \\ \vdots \\ \vdots \\ r_{\ell-2} \\ r_{\ell-1} \end{bmatrix}, \tag{1.2}$$

in which the size of the wavelet matrix $\mathbf{W}$ is $(\ell + n - 1) \times \ell$.

In the matrix–vector form of Eq. (1.2), the wavelet matrix $\mathbf{W}$ is a Toeplitz matrix, because each diagonal of $\mathbf{W}$ is a constant, $\mathbf{W}_{i,j} = \mathbf{W}_{i+1,j+1}$. Then, if we look at the wavelet matrix $\mathbf{W}$ row by row, we can see that each row of the matrix consists of the discretised wavelet samples in time-reversed order:

$$\{w_{n-1}, \cdots, w_2, w_1, w_0\}. \tag{1.3}$$

Therefore, the matrix–vector form of superposition, $\mathbf{x} = \mathbf{Wr}$, is the discretised form of 'convolution'. The corresponding continuous form for the convolution process can be expressed as

$$x(t) = \int_{-\infty}^{\infty} w(t - \tau) r(\tau) \mathrm{d}\tau = w(t) * r(t), \tag{1.4}$$

where $w(t)$ is the seismic wavelet, $r(t)$ is the subsurface reflectivity series, $*$ is the convolution operator, and $x(t)$ is the seismic trace. In this convolution process, the convolution kernel is the stationary wavelet $w(t)$, which has a constant time period, frequency, and spectral content, with respect to time variation.

A wavelet is a 'small wave', for which the time period is relatively short compared to the time duration of the reflectivity series and the resulting seismic trace. With the stationary wavelet $w(t)$, the stationary convolution process of Eq. (1.4) can be depicted as in Figure 1.2.

The stationary wavelet, that is used in Figure 1.2 for the demonstration, is the Ricker wavelet defined as (Ricker, 1953)

$$w(t) = \left[1 - 2\pi^2 f_p^2 (t - t_0)^2\right] e^{-\pi^2 f_p^2 (t-t_0)^2}, \tag{1.5}$$

where f_p is the dominant frequency (in Hz), and t_0 is the central position of the wavelet. The Ricker wavelet is symmetrical with respect to time t_0 and has a zero mean, $\int_{-\infty}^{\infty} w(t)\mathrm{d}t = 0$. Therefore, it is often known as Mexican hat wavelet in the Americas for its sombrero shape.

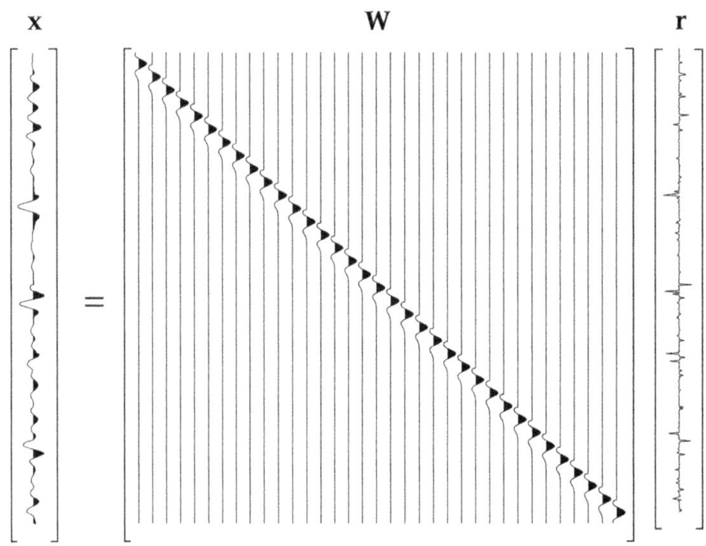

Figure 1.2 A stationary convolution process, in which a vector of seismic trace **x** is generated by the multiplication of a wavelet matrix **W** and a vector of reflectivity series **r**. Each column vector of the wavelet matrix **W** is formed by a stationary seismic wavelet $w(t)$.

1.2 Nonstationary Signals

The stationary convolution process from the previous section is an idealised model for seismic signals. In reality, however, seismic signals are nonstationary due to the dissipation effect when seismic waves propagate through the subsurface anelastic media.

The nonstationary convolution process can be expressed as follows,

$$x_a(t) = \int_{-\infty}^{\infty} w_a(t-\tau) r(\tau) \mathrm{d}\tau, \tag{1.6}$$

where $w_a(t-\tau)$ is a nonstationary seismic wavelet, in place of the stationary wavelet $w(t)$. The nonstationary seismic wavelet $w_a(t-\tau)$ evolves continually according to a dissipation model:

$$w_a(t-\tau) = w(t) * a(t,\tau) = \int_{-\infty}^{\infty} w(t-t') a(t',\tau) \mathrm{d}t', \tag{1.7}$$

where $a(t,\tau)$ is the dissipation coefficient, and acts on the idealised stationary wavelet $w(t)$.

In the convolution expression of Eq. (1.7) for the nonstationary wavelet $w_a(t-\tau)$, the dissipation coefficient $a(t,\tau)$, within which τ indicates the time dependency, is nonstationary. At any given time position τ, the dissipation coefficient $a(t,\tau)$ can be defined by

$$a(t,\tau) = \int_{-\infty}^{\infty} \mathrm{e}^{-\hat{\alpha}(f,\tau) - \mathrm{i}\hat{\beta}(f,\tau)} \mathrm{e}^{\mathrm{i}2\pi f t} \mathrm{d}f, \tag{1.8}$$

where $\mathrm{i} = \sqrt{-1}$, f is the frequency, $\hat{\alpha}(f,\tau)$ is the frequency-dependent attenuation coefficient, and $\hat{\beta}(f,\tau)$ is the associated dispersion, i.e. the phase delay of different frequency components (Futterman, 1962).

The attenuation coefficient $\hat{\alpha}(f,\tau)$ of seismic waves propagating through the subsurface anelastic media is (Wang & Guo, 2004)

$$\hat{\alpha}(f,\tau) = \int_0^{\tau} \frac{\pi f}{Q(t)} \left(\frac{f}{f_h} \right)^{-\gamma_Q(t)} \mathrm{d}t, \tag{1.9}$$

and the associated dispersion $\hat{\beta}(f,\tau)$ is

$$\hat{\beta}(f,\tau) = \int_0^\tau 2\pi f \left(\frac{f}{f_h}\right)^{-\gamma_Q(t)} \mathrm{d}t, \tag{1.10}$$

where f_h is a reference frequency,

$$\gamma_Q(t) = \frac{1}{\pi Q(t)}, \tag{1.11}$$

and $Q(t)$ is the quality factor of the subsurface anelastic media (Kolsky, 1956; Futterman, 1962; Wang, 2008). Because seismic signals have a relatively narrow frequency band, it is reasonable to assume $Q(t)$ to be frequency independent. But $Q(t)$ is time dependent, and the time t here is a proxy for geologic depth.

For numerical calculation, the nonstationary convolution process of Eq. (1.6) may also be expressed in a matrix–vector form as

$$\mathbf{x} = \mathbf{W}_a \mathbf{r}, \tag{1.12}$$

where $\mathbf{W}_a$ is the nonstationary wavelet matrix. Figure 1.3 depicts this matrix–vector form of the nonstationary convolution process.

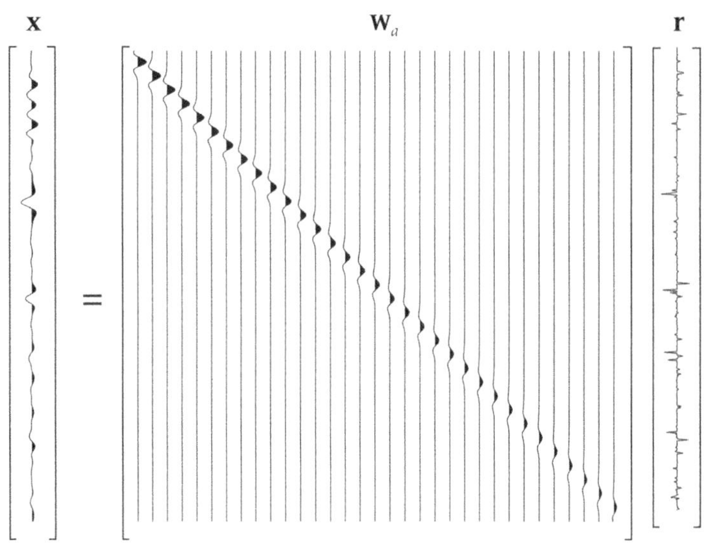

Figure 1.3 The nonstationary convolution process. The wavelet matrix $\mathbf{W}_a$ is formed by column vectors, each of which is a nonstationary seismic wavelet $w_a(t-\tau)$, and thus the seismic trace $x(t)$ is a nonstationary signal.

The essential distinction between stationary and nonstationary convolution models is the nonstationary wavelet matrix $\mathbf{W}_a$. While the example symmetrical wavelet in Figure 1.2 is a zero-phase wavelet, the form of the wavelet in Figure 1.3 changes continuously. The amplitude decreases and the phase varies with the travel time. The form of the wavelet gradually changes from symmetrical to asymmetrical.

In general, the differences between stationary and nonstationary signals can be captured in three characteristics.

(1) Time period: The time period for a stationary signal always remains constant, whereas the time period for a nonstationary signal is not constant and varies with time.
(2) Frequency: The frequency of a stationary signal remains constant throughout the process, while the frequency of a nonstationary signal changes continuously during the process.
(3) Spectral content: The spectral content of a stationary signal is constant, while the spectral content of a nonstationary signal varies continuously with respect to time.

For seismic signals, there would be several sets of frequency contents within a given time interval, and these frequency contents are likely to change dynamically with respect to travel time. Therefore, the statistical properties of nonstationary seismic signals also change with time. These statistical properties include the mean, variance, and covariance.

1.3 The Fourier Transform and the Average Properties

The Fourier transform represents a stationary signal as a stack of sinusoids, each of which has a constant frequency and a constant amplitude. Therefore, the Fourier transform is a static spectral analysis that provides only the average characteristics of a nonstationary seismic signal.

For a seismic trace $x(t)$, the Fourier transform is defined as (Bracewell, 1965)

$$\hat{x}(f) = \mathcal{F}\left[x(t)\right] = \int_{-\infty}^{\infty} x(t)\mathrm{e}^{-\mathrm{i}2\pi ft}\,\mathrm{d}t, \tag{1.13}$$

where $\hat{x}(f)$ is the frequency spectrum of $x(t)$.

The physical meaning of the Fourier transform can be understood by expressing the transform kernel in Euler's formula,

$$e^{-i2\pi ft} = \cos(2\pi ft) - i\sin(2\pi ft), \tag{1.14}$$

and rewriting the Fourier transform as follows,

$$\hat{x}(f) = \int_{-\infty}^{\infty} x(t)\cos(2\pi ft)\,dt - i\int_{-\infty}^{\infty} x(t)\sin(2\pi ft)\,dt. \tag{1.15}$$

The first integral is a cross-correlation between the seismic signal $x(t)$ and a cosine function, $\cos(2\pi ft)$, and the second integral is a cross-correlation between the seismic signal $x(t)$ and a sine function, $\sin(2\pi ft)$. These two cross-correlations reflect the content of a harmonic component f that is contained in the seismic trace.

Figure 1.4a–e demonstrate the cross-correlations between a seismic trace and the cosine function at five frequencies of 5, 20, 35, 50, and 65 Hz. The

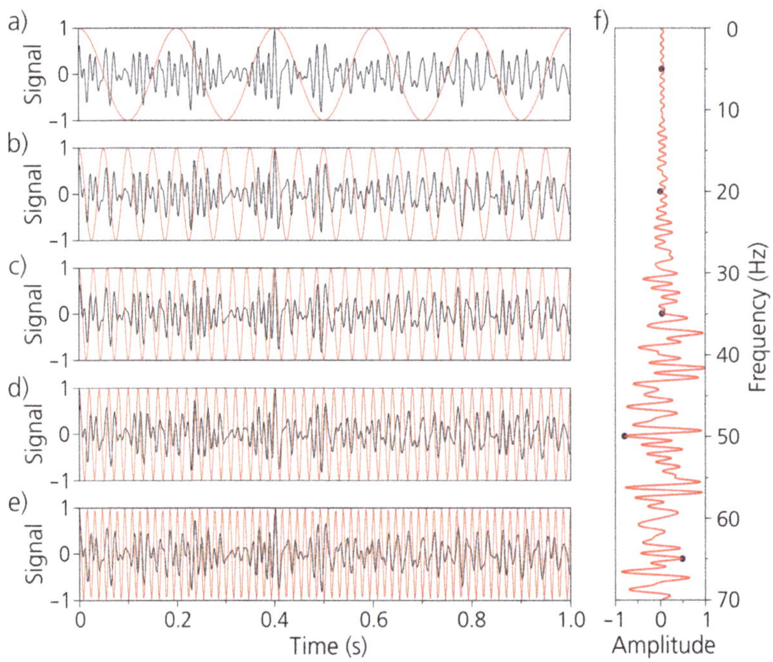

Figure 1.4 The Fourier transform is the cross-correlation between the seismic trace (black curve) and a cosine function (red curve). (a–e) The cross-correlations at five sample frequencies 5, 20, 35, 50, and 65 Hz. (f) The correlations are marked (solid circles) in the frequency spectrum.

five cross-correlation values are marked in the frequency spectrum that is displayed in the right-hand panel (Figure 1.4f).

The pair of cross-correlations is presented as a complex value, $\hat{x}(f)$, which is the frequency spectrum. The cross-correlation with the cosine function is assigned to the real part, $\hat{x}_R(f)$, and the cross-correlation with the sine function is assigned to the imaginary part, $\hat{x}_I(f)$. The magnitude of a complex value $\hat{x}(f)$ is called the amplitude at the given frequency:

$$|\hat{x}(f)| = \sqrt{\hat{x}_R^2(f) + \hat{x}_I^2(f)}. \tag{1.16}$$

The ratio of these two cross-correlations distinguishes their respective weights. Because the ratio of a sine and a cosine function is a tangent function, the inverse function of the ratio is intuitively the phase angle:

$$\theta(f) = \arg\{\hat{x}(f)\}. \tag{1.17}$$

Therefore, the frequency spectrum $\hat{x}(f)$ can be expressed in terms of the amplitude spectrum $|\hat{x}(f)|$ and the phase spectrum $\theta(f)$, as

$$\hat{x}(f) = |\hat{x}(f)| e^{i\theta(f)}. \tag{1.18}$$

The original seismic trace $x(t)$ can be reconstructed precisely from the frequency spectrum $\hat{x}(f)$, if the amplitude and phase information of all frequencies of the seismic trace is known. This is the inverse Fourier transform, defined as

$$x(t) = \mathcal{F}^{-1}[\hat{x}(f)] = \int_{-\infty}^{\infty} \hat{x}(f) e^{i2\pi ft} df. \tag{1.19}$$

The invertibility of the Fourier transform can be easily verified as follows:

$$\begin{aligned} \int_{-\infty}^{\infty} \hat{x}(f) e^{i2\pi ft} df &= \int_{-\infty}^{\infty} \left(\int_{-\infty}^{\infty} x(\tau) e^{-i2\pi f\tau} d\tau \right) e^{i2\pi ft} df \\ &= \int_{-\infty}^{\infty} x(\tau) d\tau \int_{-\infty}^{\infty} e^{i2\pi f(t-\tau)} df \\ &= \int_{-\infty}^{\infty} x(\tau) \delta(t-\tau) d\tau = x(t). \end{aligned} \tag{1.20}$$

One of the important properties of the Fourier transform is energy conservation in both the time and frequency domains (Champeney, 1987):

$$\int_{-\infty}^{\infty} |x(t)|^2 \, \mathrm{d}t = \int_{-\infty}^{\infty} |\hat{x}(f)|^2 \, \mathrm{d}f. \tag{1.21}$$

This is known as Parseval's theorem, and can be proved thus:

$$\begin{aligned}
\int_{-\infty}^{\infty} |x(t)|^2 \, \mathrm{d}t &= \int_{-\infty}^{\infty} x^*(t)x(t)\mathrm{d}t \\
&= \int_{-\infty}^{\infty} x^*(t)\left(\int_{-\infty}^{\infty} \hat{x}(f)\mathrm{e}^{\mathrm{i}2\pi ft}\mathrm{d}f\right)\mathrm{d}t \\
&= \int_{-\infty}^{\infty}\left(\int_{-\infty}^{\infty} x(t)\mathrm{e}^{-\mathrm{i}2\pi ft}\mathrm{d}t\right)^* \hat{x}(f)\mathrm{d}f \\
&= \int_{-\infty}^{\infty} \hat{x}^*(f)\hat{x}(f)\mathrm{d}f = \int_{-\infty}^{\infty} |\hat{x}(f)|^2 \, \mathrm{d}f,
\end{aligned} \tag{1.22}$$

where the superscript * represents a complex conjugate.

According to the definition of the Fourier transform, in which the time integral covers the range of $(-\infty, \infty)$, the frequency spectrum is an average quantity over the entire time duration of a seismic trace.

However, a field seismic trace is composed of signals that travel to and are reflected at different depths, and the signal properties such as amplitude and phase are time dependent. Seismic signals with such time-dependent properties are nonstationary signals.

1.4 The Analytic Signal and the Instantaneous Properties

The classical Fourier transform examines the average properties of a fairly large portion of a seismic trace, but it does not permit the local properties of a nonstationary signal to be examined. The analytic signal analysis method is a transformation method that retains local significance, and the transformation is called the Hilbert transform.

The analytic signal refers to a signal whose spectrum is zero for negative frequencies (Ville, 1948; Ackroyd, 1971). For a real-valued seismic trace

$x(t)$, the Fourier transform $\hat{x}(f)$ has Hermitian symmetry about the $f = 0$ axis:

$$\hat{x}(-f) = \hat{x}^*(f). \tag{1.23}$$

Because of Hermitian symmetry, the negative frequencies are superfluous to the Fourier transform of a real-valued signal $x(t)$. Meanwhile, it seems physically incomprehensible that a real signal has a negative frequency component at all.

The basic idea in building an analytic signal is that such a signal has no negative frequency components:

$$\hat{z}(f) = \hat{x}(f) + \operatorname{sgn}(f)\hat{x}(f), \tag{1.24}$$

where 'sgn' is the signum function,

$$\operatorname{sgn}(f) = \begin{cases} -1, & f < 0, \\ 0, & f = 0, \\ 1, & f > 0. \end{cases} \tag{1.25}$$

The frequency spectrum $\hat{z}(f)$ contains only the non-negative frequency components of $\hat{x}(f)$. Then the analytic signal $z(t)$ in the time domain is obtained by the inverse Fourier transform of $\hat{z}(f)$:

$$\begin{aligned} z(t) &= \mathcal{F}^{-1}\left[\hat{x}(f) + \operatorname{sgn}(f)\hat{x}(f)\right] \\ &= \mathcal{F}^{-1}\left[\hat{x}(f)\right] + \mathcal{F}^{-1}\left[\operatorname{sgn}(f)\right] * \mathcal{F}^{-1}\left[\hat{x}(f)\right] \\ &= x(t) + \text{p.v.}\left(\frac{\mathrm{i}}{\pi t}\right) * x(t) \\ &= x(t) + \mathrm{i}\mathcal{H}[x(t)] \\ &= x(t) + \mathrm{i}\,y(t), \end{aligned} \tag{1.26}$$

where $y(t) = \mathcal{H}[x(t)]$ is the Hilbert transform of $x(t)$, and 'p.v.' denotes the Cauchy principal value (Hilbert, 1912; Hahn, 1996).

According to Eq. (1.26), the analytic signal $z(t)$ is a complex signal, in which the real part $x(t)$ and the imaginary part $y(t)$ are related to each other by the Hilbert transform, which is a time-domain convolution:

$$\mathcal{H}[x(t)] = \text{p.v.}\left(\frac{1}{\pi t}\right) * x(t) = \frac{1}{\pi}\text{p.v.}\int_{-\infty}^{\infty} \frac{x(\tau)}{t-\tau}\mathrm{d}\tau$$

$$= \frac{1}{\pi}\lim_{t\to\tau^-}\int_{-\infty}^{t-\tau} \frac{x(\tau)}{t-\tau}\mathrm{d}\tau + \frac{1}{\pi}\lim_{t\to\tau^+}\int_{t-\tau}^{\infty} \frac{x(\tau)}{t-\tau}\mathrm{d}\tau. \tag{1.27}$$

The physical meaning of the Hilbert transform is a $\pm\pi/2$ phase rotation. According to Eq. (1.26), $\mathcal{F}^{-1}[\mathrm{sgn}(f)\hat{x}(f)] = \mathrm{i}y(t)$. Therefore, for the Hilbert transform $y(t) = \mathcal{H}[x(t)]$, the Fourier transform is

$$\hat{y}(f) = \mathcal{F}[y(t)] = -\mathrm{i}\,\mathrm{sgn}(f)\hat{x}(f). \tag{1.28}$$

Because $-\mathrm{i}\,\mathrm{sgn}(f) = \mathrm{e}^{\pm\mathrm{i}\pi/2}$, for $f \neq 0$, the Hilbert transform imparts a phase shift of $+\pi/2$ to every negative frequency component of the original signal and a phase shift of $-\pi/2$ to every positive frequency component of the original signal. Because of the $\pi/2$ phase offset, $x(t)$ and $y(t) = \mathcal{H}[x(t)]$ are quadrature counterparties.

The analytic signal $z(t)$, that is a complex function of time t, can be written as (Taner *et al.*, 1979)

$$z(t) = a(t)\mathrm{e}^{\mathrm{i}\theta(t)}, \tag{1.29}$$

where $a(t)$ is the instantaneous amplitude of the original signal $x(t)$,

$$a(t) = \sqrt{x^2(t) + y^2(t)}, \tag{1.30}$$

and $\theta(t)$ is the instantaneous phase of the original signal $x(t)$,

$$\theta(t) = \arg\{z(t)\} = \tan^{-1}\left(\frac{y(t)}{x(t)}\right). \tag{1.31}$$

In addition to the instantaneous amplitude and the instantaneous phase, another characteristic feature of the nonstationary signal is the instantaneous frequency, which changes constantly with time.

According to Eq. (1.29), the instantaneous frequency is the frequency of a sinusoid that best fits the complex signal locally. Therefore, the Fourier transform of $z(t)$ becomes

$$\hat{z}(f) = \int_{-\infty}^{\infty} z(t)\mathrm{e}^{-\mathrm{i}2\pi ft}\mathrm{d}t = \int_{-\infty}^{\infty} a(t)\mathrm{e}^{\mathrm{i}[\theta(t)-2\pi ft]}\,\mathrm{d}t. \tag{1.32}$$

At time t, the phase is stationary,

$$\frac{\mathrm{d}}{\mathrm{d}t}\left[\theta(t) - 2\pi ft\right] = 0. \tag{1.33}$$

This stationary phase condition leads to the definition of the instantaneous frequency:

$$f_{\mathrm{inst}}(t) = \frac{1}{2\pi}\frac{\mathrm{d}\theta(t)}{\mathrm{d}t}. \tag{1.34}$$

The physical meaning of the instantaneous frequency (Ackroyd, 1970) can be understood as the centroid of the instantaneous spectrum at time t:

$$p(t, f) = z(t)\hat{z}^*(f)\mathrm{e}^{-\mathrm{i}2\pi ft}. \tag{1.35}$$

This is a time–frequency spectrum, representing the frequency spectrum of signal $z(t)$ at time t (Ville, 1948).

1.5 Computation of the Instantaneous Frequency

The instantaneous frequency $f_{\mathrm{inst}}(t)$ is defined in Eq. (1.34) as the rate of change of the instantaneous phase $\theta(t)$. However, this definition cannot be used directly to calculate the instantaneous frequency. The arctangent calculation of Eq. (1.31) provides only the principal value of the instantaneous phase $\theta(t)$, which must be unwrapped to remove the 2π phase jumps and form a continuous function for the calculation of the instantaneous frequency. Even though, the phase unwrapping process cannot iron out the $\pm\pi$ phase jumps caused by spikes in the noisy seismic trace (Wang, 1998, 2000) and, thus, cannot produce an idealised continuous phase function $\theta(t)$ for the time differential.

A convenient way of computing the instantaneous frequency $f_{\mathrm{inst}}(t)$ is to compute the time differentials, $\mathrm{d}x(t)/\mathrm{d}t$ and $\mathrm{d}y(t)/\mathrm{d}t$. Rewriting the arctangent function of Eq. (1.31) as

$$\tan\theta(t) = \frac{y(t)}{x(t)} \tag{1.36}$$

and differentiating both sides with respect to time t, we have

$$\frac{1}{\cos^2\theta(t)}\frac{\mathrm{d}\theta(t)}{\mathrm{d}t} = \frac{x(t)\dfrac{\mathrm{d}y(t)}{\mathrm{d}t} - \dfrac{\mathrm{d}x(t)}{\mathrm{d}t}y(t)}{x^2(t)}. \tag{1.37}$$

If $a^2(t) = x^2(t) + y^2(t) \neq 0$, substituting

$$\cos^2\theta(t) = \frac{x^2(t)}{x^2(t) + y^2(t)}, \tag{1.38}$$

we obtain the formula of the instantaneous frequency $f_{\text{inst}}(t)$, as

$$f_{\text{inst}}(t) = \frac{1}{2\pi} \frac{x(t)\frac{\mathrm{d}y(t)}{\mathrm{d}t} - \frac{\mathrm{d}x(t)}{\mathrm{d}t}y(t)}{x^2(t) + y^2(t)}. \tag{1.39}$$

Note that, according to the definition of the analytic signal $z(t)$ for a real signal $x(t)$ (Eq. 1.24), the instantaneous frequency must be non-negative, $f_{\text{inst}}(t) \geq 0$.

The noisy instantaneous frequency can be further smoothed by a weighted averaging:

$$\tilde{f}_{\text{inst}}(t) = \frac{\rho \int_{-\infty}^{\infty} L(t-\tau)a(\tau)f_{\text{inst}}(\tau)\mathrm{d}\tau}{\int_{-\infty}^{\infty} L(t-\tau)a(\tau)\mathrm{d}\tau}, \tag{1.40}$$

where $L(t)$ is a low-pass filter along the time axis, and ρ is a scaling factor, which is inversely proportional to the size of the low-pass filter. In the averaging formula of Eq. (1.40), the instantaneous amplitude $a(t)$ is used as a weighting function to the instantaneous frequency $f_{\text{inst}}(t)$. The averaging of the instantaneous frequency $f_{\text{inst}}(t)$ over a time window $L(t)$ ensures that the smoothed frequency $\tilde{f}_{\text{inst}}(t)$ is no greater than the Nyquist frequency of a discretised data series (Barnes, 1992).

Figure 1.5a shows a synthetic signal, composed of energy-normalised Ricker wavelets, defined as

$$w(t) = 2\sqrt{\frac{f_p\sqrt{2\pi}}{3}}(1 - 2\pi^2 f_p^2 t^2)e^{-\pi^2 f_p^2 t^2}, \tag{1.41}$$

where f_p is the peak frequency. The peak frequencies of the five wavelets are 60, 50, 40, 30, and 20 Hz, respectively. The energy normalisation mimics an idealised situation with no energy attenuation during wave propagation. It can be proved that the energy of this Ricker wavelet (Eq. 1.41) is unity:

$$\int_{-\infty}^{\infty} |w(t)|^2 \mathrm{d}t = 1. \tag{1.42}$$

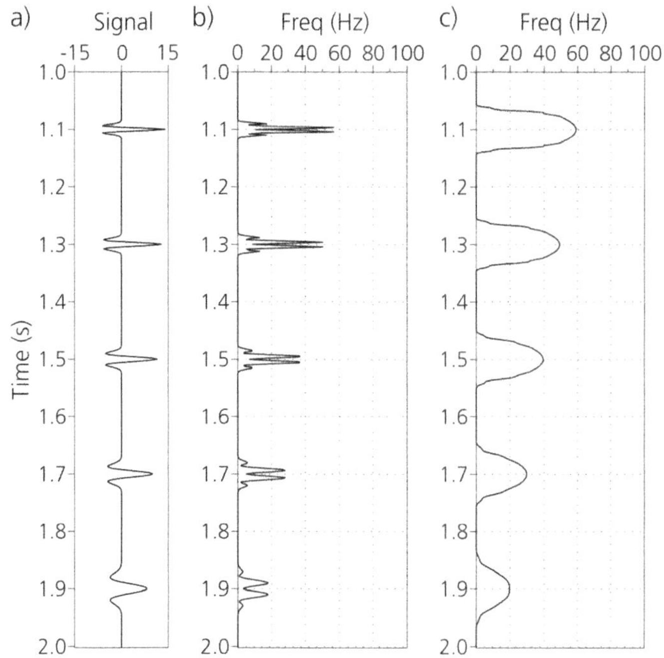

Figure 1.5 The concept of the instantaneous frequency. (a) A synthetic trace composed of energy-normalised Ricker wavelets. (b) The instantaneous frequency $f_{\text{inst}}(t)$. (c) The smoothed instantaneous frequency $\tilde{f}_{\text{inst}}(t)$.

The proof involves a Gaussian integral (Appendix A):

$$\int_0^\infty t^n e^{-at^2} dt = \frac{1}{2a^{(n+1)/2}} \Gamma\left(\frac{n+1}{2}\right), \tag{1.43}$$

where n is a positive integer, and $\Gamma(u)$ is the gamma function. In the definition of $\Gamma(u)$, the real part of variable u is positive, $\text{Re}(u) > 0$.

Figure 1.5b shows the instantaneous frequency $f_{\text{inst}}(t)$ computed through the time differentials of the original signal $x(t)$ and the Hilbert transform $y(t) = \mathcal{H}[x(t)]$. The computation result contains strong numerical noise.

Figure 1.5c displays the smoothed frequency $\tilde{f}_{\text{inst}}(t)$. The low-pass filter $L(t)$ that is used is a Gaussian function with the Gaussian spread parameter $\sigma_t = 4\Delta t$, where Δt is the sampling rate in time. Numerical experiments suggest that the scaling factor in Eq. (1.40) can be set as $\rho = 2$ for the low-pass filter $L(t)$ with $\sigma_t \geq 4\Delta t$.

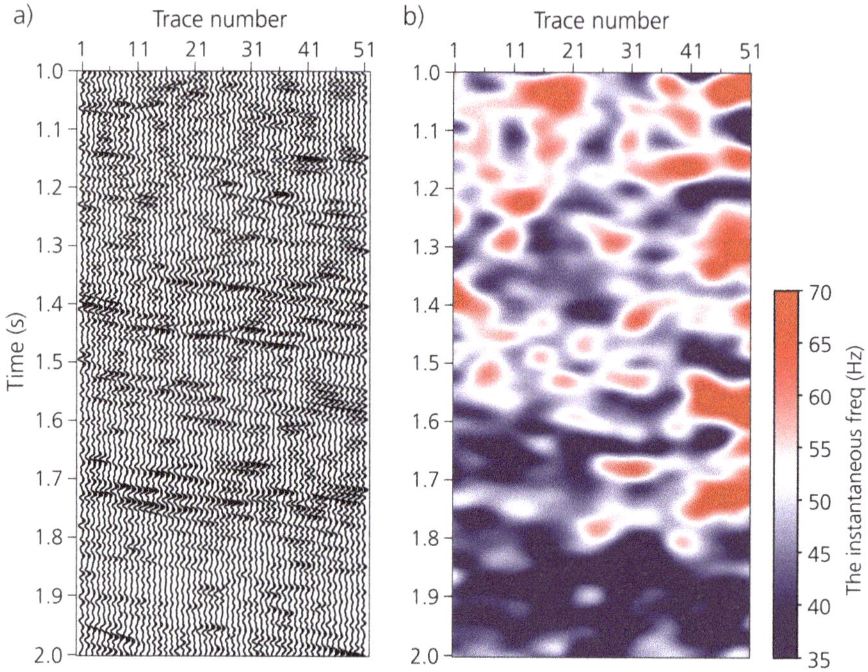

Figure 1.6 (a) A two-dimensional seismic profile. (b) The instantaneous frequency profile, which is smoothed in both the time and distance directions.

Figure 1.6 is an example of seismic profile $x(t, d)$, and the corresponding profile of instantaneous frequencies $\tilde{f}_{\text{inst}}(t, d)$. The instantaneous frequency function is calculated and is smoothed trace by trace. In addition, it is also smoothed along the horizontal direction, by using a Gaussian filter with the spread parameter $\sigma_d = \sigma_t \Delta d / (2\Delta t)$, where Δd is the trace interval.

While for nonstationary seismic signals the instantaneous frequency and the instantaneous amplitude can be examined with the analytic signal analysis method described above, the spectral contents, which also vary in time, have to be examined in the time–frequency plane. Therefore, different methods for analysing the time–frequency spectra of nonstationary seismic signals are presented in this book.

1.6 Two Groups of Time–Frequency Analysis Methods

Time–frequency analysis can be considered as a generalisation of the classical Fourier transform of a stationary signal, and applied to the spectral analysis of a nonstationary signal. All methods of time–frequency analysis can be implemented by the Fourier transform, but they differ in the construction of the transform kernel. Based on the difference in the transformation kernel, I divide the different methods of time–frequency analysis into two groups.

The first group of methods is *the Gabor transform-type method*. This group of methods applies various time-window functions to the seismic signal, to form a signal segment, prior to initiating the Fourier transform. In the short-time Fourier transform (STFT), a segment is taken directly from the original signal for the Fourier transform, assuming that the signal segment can be treated as stationary on a short-time basis. The signal segmentation process can be understood as the application of a rectangular window. The application of different time-window functions results in

(1) the Gabor transform (Chapter 2),
(2) the continuous wavelet transform (Chapter 3),
(3) the S transform (Chapter 4), and
(4) the W transform (Chapter 5).

I name this group of methods after Dennis Gabor (1946a, 1946b, 1946c), the pioneer of the STFT method.

The second group of methods is *the energy density distribution method*. The formulation of the transform kernel involves manipulation of the original signal, and the time–frequency representation is the energy density distribution. This group of methods includes

(1) the Wigner–Ville distribution (WVD) (Chapter 6),
(2) the WVD based on the matching pursuit (Chapter 7), and
(3) the local power spectra with multiple windows (Chapter 8).

The kernel in the WVD is the instantaneous autocorrelation of the signal, and not the original signal. The kernel in the WVD based on matching pursuit is the instantaneous autocorrelation of a wavelet, extracted from the original signal by matching pursuit. The local power spectral analysis method manipulates the transform kernel with multiple orthogonal windows. It applies multiple windows in parallel to the same signal segment, and the combined effect of these windows is optimal in retaining the local spectral information of the signal segment.

For the Gabor transform-type method, there are two important features:

(1) The time–frequency spectra produced by the Gabor transform-type method are complex, and the complex spectra consist of both the amplitude spectrum and the phase spectrum.

(2) Consequently, the original signal can be reconstructed by an inverse transform applied to the time–frequency spectrum.

Therefore, the time–frequency spectra can be used in signal processing. If applying a two-dimensional filter to the time–frequency spectrum, the inverse transform produces a filtered signal.

In contrast, two features of the energy density distribution method are:

(1) The energy density distribution in the time–frequency plane is real, and the real spectrum contains no phase information.
(2) Consequently, the original time signal cannot be recovered from the energy density distribution.

The energy density distribution in the time–frequency plane can be used as a seismic attribute for geophysical characterisation.

2

The Gabor Transform

The development of time–frequency spectral analysis goes back to the classical Fourier transform. The classical Fourier transform analyses the average properties of a seismic trace and does not permit examination of the local properties of the nonstationary signal. A simple method for analysing local spectral properties is to take a short segment from the seismic trace and perform the classical Fourier transform over the segment. This basic procedure is performed sequentially by following the time axis of the seismic trace.

2.1 Short-time Fourier Transform

In the basic procedure called the short-time Fourier transform (STFT), a rectangular window is used to extract a short segment from the signal.

The first step is to multiply the original seismic trace by a window function. At a given time t, the windowed signal segment is

$$s(t,\tau) = w(\tau - t)x(\tau) = w(t - \tau)x(\tau), \tag{2.1}$$

where $x(t)$ is the original seismic trace, $w(t)$ is a symmetric time window function, and $s(t,\tau)$ is the windowed signal segment centred at time t. The second step is to perform the Fourier transform over the signal segment $s(t,\tau)$, with respect to time τ,

$$S_{\mathrm{STFT}}(t,f) = \int_{-\infty}^{\infty} s(t,\tau)e^{-i2\pi f\tau}d\tau, \tag{2.2}$$

Time–Frequency Analysis of Seismic Signals, First Edition. Yanghua Wang.

where $S_{\mathrm{STFT}}(t,f)$ is the time–frequency spectrum produced by the STFT. Combining these two steps, the STFT spectrum can be expressed as follows,

$$S_{\mathrm{STFT}}(t,f) = \int_{-\infty}^{\infty} w(t-\tau)x(\tau)\mathrm{e}^{-\mathrm{i}2\pi f\tau}\mathrm{d}\tau. \tag{2.3}$$

The rectangular window used in classical STFT is called the boxcar window function:

$$w(t) = \begin{cases} \dfrac{1}{\sqrt{2T}}, & \text{for } -T \leq t \leq T, \\ 0, & \text{otherwise,} \end{cases} \tag{2.4}$$

where T is the (half-)length of the window. The Fourier transform of the boxcar window function is

$$\hat{w}(f) = \frac{1}{\sqrt{2T}} \int_{-T}^{T} \mathrm{e}^{-\mathrm{i}2\pi ft}\mathrm{d}t = \sqrt{2T}\left(\frac{\sin(2\pi fT)}{2\pi fT}\right). \tag{2.5}$$

The fraction in the brackets is a sinc function, $\mathrm{sinc}(u) = \sin(u)/u$.

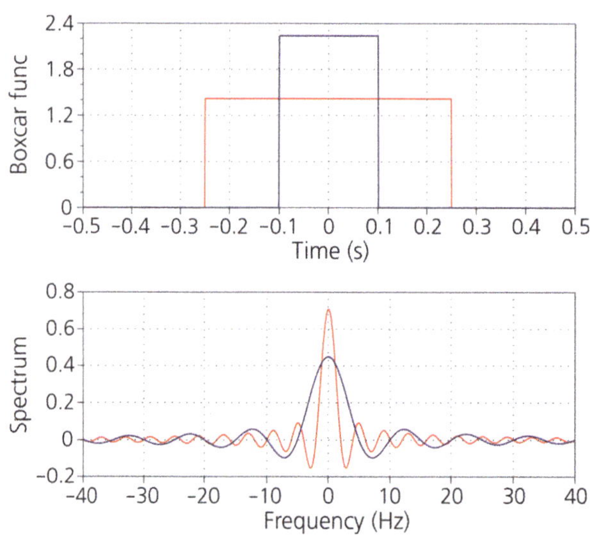

Figure 2.1 Two boxcar window functions and their corresponding frequency spectra. A narrower window function produces a wider frequency spectrum, and a wider window function produces a more restricted frequency spectrum.

The energy of the window function is normalised to unity,

$$\int_{-\infty}^{\infty} |w(t)|^2 \, \mathrm{d}t = \int_{-\infty}^{\infty} |\hat{w}(f)|^2 \, \mathrm{d}f = 1. \tag{2.6}$$

When a seismic trace is multiplied by this normalised window function, the spectral energy is not affected.

With the boxcar window function, there is only one parameter T that is adjustable. We can understand the influence of this parameter by looking at the frequency spectrum of the boxcar window function. Figure 2.1 compares two boxcar window functions, defined with the parameter $T =$ 0.1 s and 0.25 s, respectively. The corresponding spectra show that a narrower window function produces a wider frequency spectrum, and a wider window function produces a more restricted frequency spectrum.

Consider the multiplication of the seismic trace by a window function in the time domain as a convolution in the frequency domain. If the window function is too short and the frequency spectrum too wide, all sharp local spectra of the signal components will be spread out. This means that a narrower window function has poor spectral resolution. If the window function is long and the frequency spectrum is narrow, it has better spectral resolution and does not significantly change the signal spectrum during convolution in the frequency domain. However, the spectral feature is also not localised, because the observation window is too large, and the spectrum mixes different signal components in time. We will discuss this trade-off in time–frequency resolution in Section 2.4.

The spectra in Figure 2.1 also reveal a problem associated with the boxcar window function. The frequencies around the central peak of the amplitude spectrum represent a low-frequency passband. Apart from this passband, there are strong oscillations along the frequency axis (Allen, 1977; Hall, 2006). These oscillations indicate positively or negatively weighted spectral leakage outside the low-frequency passband.

To reduce the spectral leakage, the boxcar window function can be tapered. A tapered boxcar window function can be defined as

$$w(t) = \frac{1}{\sqrt{2T + \sqrt{\pi}\sigma}} \begin{cases} \mathrm{e}^{-\frac{(t+T)^2}{2\sigma^2}}, & \text{for } -\infty < t < -T, \\ 1, & \text{for } -T \le t \le T, \\ \mathrm{e}^{-\frac{(t-T)^2}{2\sigma^2}}, & \text{for } T < t < \infty, \end{cases} \tag{2.7}$$

where σ controls the width of the taper. Here, the taper is a Gaussian function in which the standard deviation σ is the spread parameter.

The corresponding Fourier transform of the tapered boxcar window function is

$$\hat{w}(f) = \frac{2T}{\sqrt{2T+\sqrt{\pi}\sigma}}\left(\frac{\sin(2\pi fT)}{2\pi fT}\right) + \frac{\sqrt{2\pi}\sigma}{\sqrt{2T+\sqrt{\pi}\sigma}} e^{-2\pi^2 f^2 \sigma^2}\left[\cos(2\pi fT) - \sin(2\pi fT)\operatorname{erfi}(\sqrt{2}\pi f\sigma)\right], \tag{2.8}$$

where erfi(z) is the imaginary error function:

$$\operatorname{erfi}(z) = \frac{2}{\sqrt{\pi}}\int_0^z e^{t^2}\,dt. \tag{2.9}$$

If we look at the definition of Eq. (2.9), we can see that if z is real, erfi(z) is real, despite the name 'imaginary'. The imaginary error function is related to the Gaussian error function as follows (Appendix A):

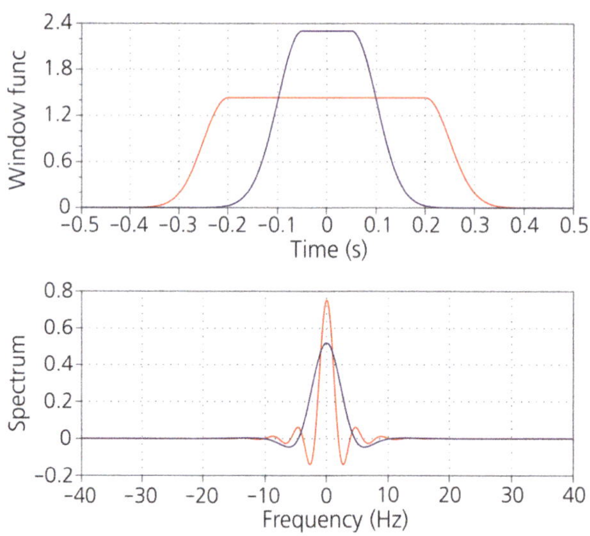

Figure 2.2 Two tapered boxcar window functions and their corresponding spectra. Tapering a boxcar window function can effectively reduce the spectral oscillations of the boxcar window function, as shown in Figure 2.1.

$$\mathrm{erfi}(z) = -\mathrm{i}\,\mathrm{erf}(\mathrm{i}z), \tag{2.10}$$

where $\mathrm{erf}(z)$ is the Gaussian error function,

$$\mathrm{erf}(z) = \frac{2}{\sqrt{\pi}} \int_0^z \mathrm{e}^{-t^2} \mathrm{d}t, \tag{2.11}$$

which is an integral of the truncated Gaussian function. In contrast, $\mathrm{erfi}(z)$ is an integral of the inverse of the truncated Gaussian function.

The Fourier transform in Eq. (2.8) is derived in Appendix B.1. Note that in the numerical calculation of Eq. (2.8), in order to avoid arithmetic overflow caused by e^{t^2} in the imaginary error function (Eq. 2.9), the imaginary error function $\mathrm{erfi}(\sqrt{2}\pi f \sigma)$ should be considered together with the exponential factor $\mathrm{e}^{-2\pi^2 f^2 \sigma^2}$.

Figure 2.2 shows two boxcar window functions with tapering and the corresponding frequency spectra. The two tapered boxcar window functions are defined with the parameters $(T, \sigma) = (0.2\,\mathrm{s},\ 0.05\,\mathrm{s})$ and $(0.05\,\mathrm{s},\ 0.05\ \mathrm{s})$, respectively. Their spectra clearly show that tapering a boxcar window function can effectively reduce the oscillations in the frequency spectrum, although there is still some spectral leakage immediately adjacent to the low-frequency passband.

2.2 The Gabor Transform

The Gabor transform[1] is a generalised form of the STFT. The STFT method described in the previous section suffers from spectral leakage when a boxcar window function is used, whether the window function is tapered or not. The Gabor transform uses a time window function without such spectral leakage.

To mitigate the spectral leakage in the STFT, one could simply set $T = 0$ in the tapered boxcar window function (Eq. 2.7), that is, defining the window function in a Gaussian form as

$$w(t) = \frac{1}{\sqrt[4]{\pi\sigma^2}} \mathrm{e}^{-\frac{t^2}{2\sigma^2}}. \tag{2.12}$$

---

[1] The Gabor transform, after Dennis Gabor (1900–1979), a Nobel Laureate of Imperial College London (Gabor, 1946a).

Then, the Gabor transform (GT) can be expressed as

$$S_{GT}(t,f) = \frac{1}{\sqrt[4]{\pi\sigma^2}} \int_{-\infty}^{\infty} e^{-\frac{(t-\tau)^2}{2\sigma^2}} x(\tau) e^{-i2\pi f\tau} d\tau. \tag{2.13}$$

Applying the Gaussian window function to a seismic trace means that the highest weight is applied to the signal only at the time position being analysed. However, the energy of the Gaussian window function is normalised to unity, so that the window function does not affect the spectral energy, in accordance with Parseval's theorem (Eq. 1.21).

The frequency spectrum of the Gaussian window function is

$$\hat{w}(f) = \sqrt[4]{4\pi\sigma^2}\, e^{-2\pi^2 f^2 \sigma^2}. \tag{2.14}$$

The frequency spectrum is also in a Gaussian form which, as shown in Figure 2.3, completely mitigates the spectral leakage occurred in the boxcar window function, either with or without tapering.

In the Gaussian window function, the spread parameter σ (in seconds) controls the width of the Gaussian envelope. The (half-)width at half the maximum magnitude is

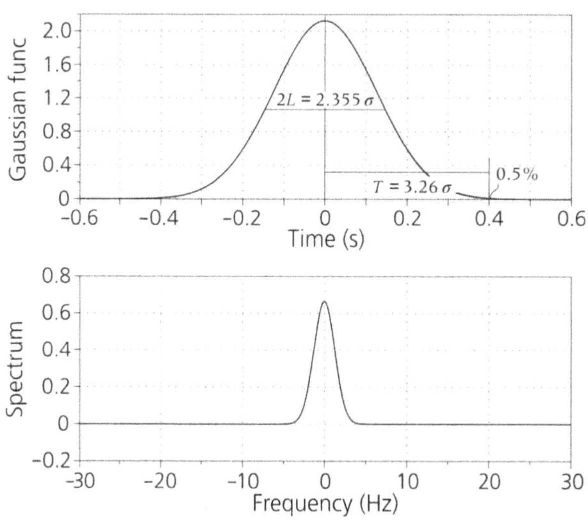

Figure 2.3 The Gaussian function (with $\sigma = 0.125$s), and the corresponding spectrum. The spread parameter σ determines the width of the Gaussian envelope. In practice, the Gaussian window can be truncated at $T = 3.26\sigma$, where the magnitude of the envelope is 0.5% of the maximum magnitude, or –46 dB.

$$L = \sqrt{2\ln 2}\sigma \approx 1.177\sigma. \qquad (2.15)$$

This width parameter represents the temporal resolution of the window function. Consequently, this spread parameter also controls the spectral resolution of the signal, since the frequency spectrum $\hat{w}(f)$ of the window function is a convolution operator in the frequency domain applied to the frequency spectrum of a signal segment.

The Gaussian window function is not really a 'window function' because the function takes infinite time. Such an infinite duration is impractical for implementation because even in the extreme case with an infinite window duration, a seismic trace always has a finite duration. Therefore, the idealised Gaussian function must be truncated to obtain a true window function.

If the Gaussian envelope is truncated at a level of significance, for instance, at 0.5% of the maximum magnitude or −46 dB, the (half-)length of the time-domain Gaussian envelope is

$$T = \sqrt{2\times(3\ln 2 + 2\ln 5)}\sigma \approx 3.26\sigma. \qquad (2.16)$$

Then, the truncated Gaussian window function is defined as

$$w(t) = \frac{1}{\sqrt[4]{\pi\sigma^2}} \begin{cases} e^{-\frac{t^2}{2\sigma^2}}, & \text{for } -T \le t \le T, \\ 0, & \text{otherwise,} \end{cases} \qquad (2.17)$$

and the Fourier transform is

$$\hat{w}(f) = \sqrt[4]{4\pi\sigma^2}\, e^{-2\pi^2 f^2 \sigma^2}\, \mathrm{Re}\left[\mathrm{erf}\left(\frac{T}{\sqrt{2}\sigma} + i\sqrt{2}\pi f\sigma\right)\right]. \qquad (2.18)$$

where $\mathrm{erf}(z)$ is the Gaussian error function that is defined in Eq. (2.11). This Fourier transform is derived in Appendix B.2.

The variable z of the Gaussian error function $\mathrm{erf}(z)$ in Eq. (2.18) is a complex number. Considering a complex integral as a path integral,

$$\int_C e^{-t^2} dt = \int_C e^{-(u+iv)^2} du + i e^{-(u+iv)^2} dv, \qquad (2.19)$$

where C is an arbitrary path from 0 to z, and $t = u + iv$. Denoting $P = e^{-(u+iv)^2}$, $Q = ie^{-(u+iv)^2}$, we can prove $\partial P/\partial v = \partial Q/\partial u$. According to Green's theorem, the path integral of Eq. (2.19) is independent from the path selection. Therefore, for a complex number $z = z_R + iz_I$, the Gaussian error function $\mathrm{erf}(z)$ can be expressed explicitly as

$$\mathrm{erf}(z_R + \mathrm{i}z_I) = \frac{2}{\sqrt{\pi}} \int_0^{z_R+\mathrm{i}z_I} \mathrm{e}^{-t^2} \mathrm{d}t = \frac{2}{\sqrt{\pi}} \int_0^{z_R} \mathrm{e}^{-(u+\mathrm{i}0)^2} \mathrm{d}u + \mathrm{i}\frac{2}{\sqrt{\pi}} \int_0^{z_I} \mathrm{e}^{-(z_R+\mathrm{i}v)^2} \mathrm{d}v$$

$$= \frac{2}{\sqrt{\pi}} \int_0^{z_R} \mathrm{e}^{-u^2} \mathrm{d}u + \mathrm{i}\mathrm{e}^{-z_R^2} \frac{2}{\sqrt{\pi}} \int_0^{z_I} \mathrm{e}^{-\mathrm{i}2z_R v} \mathrm{e}^{v^2} \mathrm{d}v. \tag{2.20}$$

Here, the path integral is performed along two line-segments, $(0,0) \to (z_R, 0)$, and $(z_R, 0) \to (z_R, z_I)$ (Zaghloul & Ali, 2012).

Truncating the Gaussian window function potentially leads to spectral leakage, which we will discuss later. There are several window functions that have similar shapes to the truncated Gaussian window, and they all suffer from spectral leakage to some degree.

2.3 The Cosine Function Windows

As explained in the previous two sections, one purpose of developing a window function as a multiplicative weighting function applied to the signal is to reduce discontinuity at the boundaries of the signal segment. This purpose can be achieved by fitting as many derivative orders of the weighted segment as possible at the boundaries (Harris, 1978).

A popular group of window functions is a stack of weighted cosine functions over a finite time duration (Nuttall, 1981):

$$w(t) = \begin{cases} \frac{1}{r}\sum_{k=0}^{K-1} c_k \cos\frac{k\pi t}{T}, & \text{for } -T \le t \le T, \\ 0, & \text{otherwise,} \end{cases} \tag{2.21}$$

where K is the number of cosine terms with which to define a cosine-function window, and r is the normalisation factor,

$$r = \sqrt{\left(c_0^2 + \sum_{k=0}^{K-1} c_k^2 \right) T}, \tag{2.22}$$

which makes the energy of the cosine-function window unity. The Fourier transform of a weighted cosine-function window is (Appendix B.3)

$$\hat{w}(f) = \sqrt{\frac{T}{c_0^2 + \sum_{k=0}^{K-1} c_k^2}} \sum_{k=0}^{K-1} c_k \left(\frac{\sin(2\pi fT - k\pi)}{2\pi fT - k\pi} + \frac{\sin(2\pi fT + k\pi)}{2\pi fT + k\pi} \right). \tag{2.23}$$

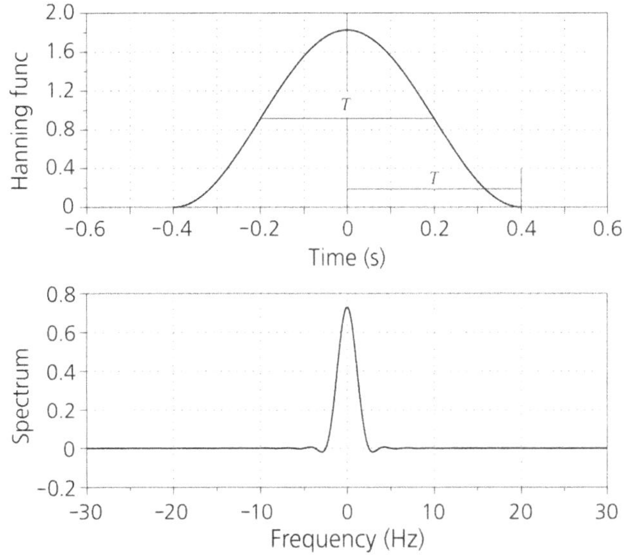

Figure 2.4 The Hanning window function (with $T = 0.4$ s) and the corresponding frequency spectrum. The (half-)time length of the envelope T also determines the width of the window function, measured at half the maximum magnitude.

The simplest weighted cosine-function window is composed of only two terms, $c_0 \neq 0$, $c_1 \neq 0$, for $-T \leq t \leq T$. While the (half-)length of the time-domain envelope is T, because $w(t = \pm T) = 0$, the (half-)width of the window function is

$$L = \frac{T}{\pi} \cos^{-1}\left(\frac{1}{2c_1} - \frac{c_0}{c_1} \right), \tag{2.24}$$

which is measured at half the maximum magnitude.

If $c_0 = c_1 = 0.5$ (Hann, 1903), it is called a Hanning window [2]. As shown in Figure 2.4, the width of the window function is $2L = T$, at half the maximum magnitude.

The frequency spectrum of the Hanning window is similar to the spectrum of the Gaussian window function, but there is smaller spectral leakage evident in the frequency spectrum of the Hanning window.

---

[2] The Hanning window, after the Austrian meteorologist Julius Ferdinand von Hann (1839–1921).

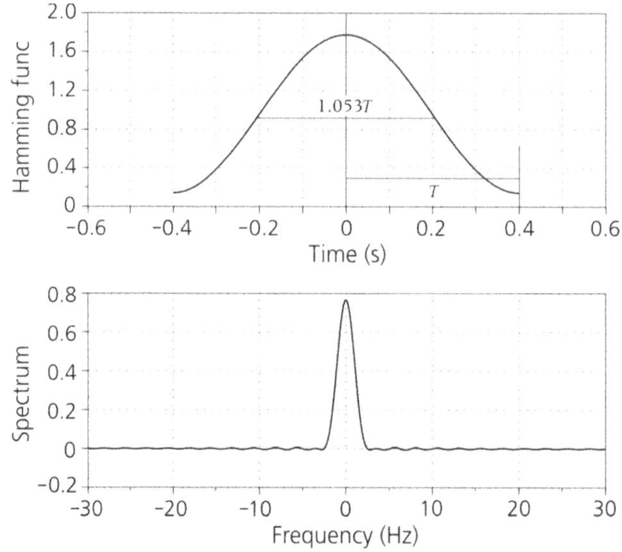

Figure 2.5 The Hamming window function (with $T = 0.4$ s) and the corresponding frequency spectrum. The width of the window function at half the maximum magnitude is $2L \approx 1.053T$.

The largest sidelobe can be further reduced, as shown in Figure 2.5, if we set $(c_0, c_1) = (0.53836, 0.46164)$ (Tukey & Hamming, 1949). This is called a Hamming window[3]. The width of the Hamming window function is $2L \approx 1.053T$, at half the maximum magnitude.

To further suppress the sidelobes, a three-term or a four-term cosine-function window can be used. The (half-)width of the three-term window is

$$L = \frac{T}{\pi}\cos^{-1}\left(\frac{\sqrt{c_1^2 - 8c_2^2 + 4c_2(1 - 2c_0)} - c_1}{4c_2}\right), \tag{2.25}$$

measured at half the maximum magnitude. The Blackman window is a sample three-term cosine-function window, defined by the following three coefficients (Blackman & Tukey, 1958; Harris, 1978; Albrecht, 2001):

$$(c_0, c_1, c_2) = (0.42438, 0.49734, 0.07828). \tag{2.26}$$

[3] The Hamming window, after an American mathematician Richard Wesley Hamming (1915–1998).

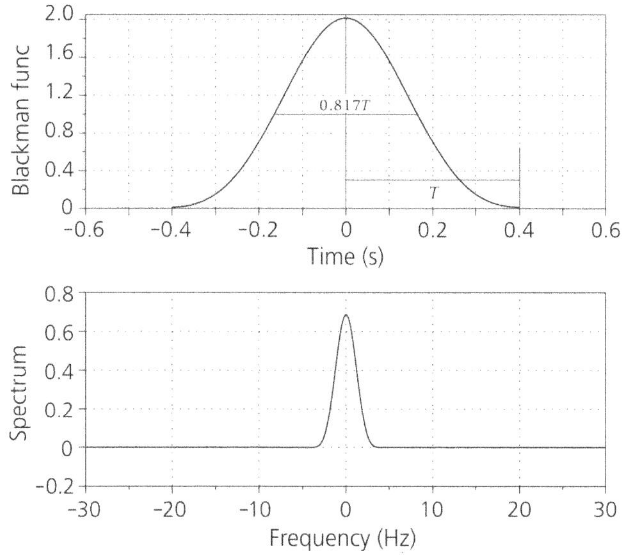

Figure 2.6 The Blackman window function (with $T = 0.4\,\text{s}$) and the corresponding frequency spectrum. The width of the window function at half the maximum magnitude is $2L \approx 0.817T$.

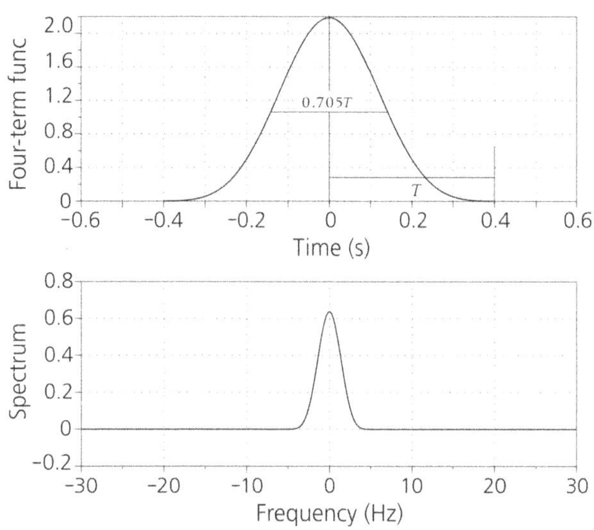

Figure 2.7 A Blackman–Harris window function with four terms (with $T = 0.4\,\text{s}$) and the corresponding frequency spectrum. The width of the window function at half the maximum magnitude is $2L \approx 0.705T$.

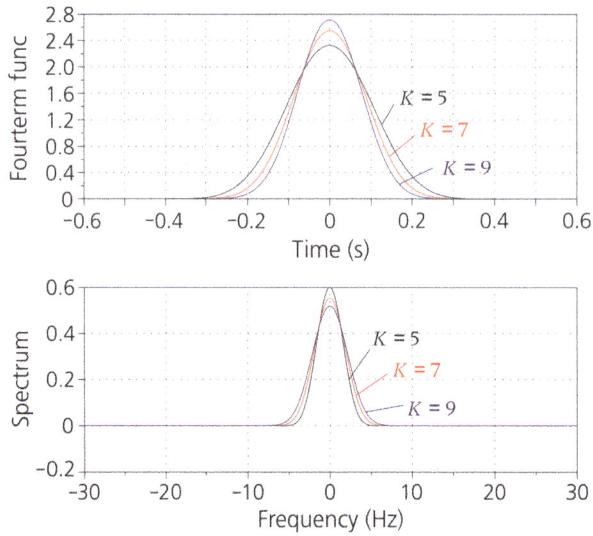

Figure 2.8 The Blackman–Harris window functions, with $K = (5, 7, 9)$ and $T = 0.4$ s, and the corresponding frequency spectra. The width of the window function becomes somewhat narrower as the number of terms increases, and the corresponding spectrum becomes somewhat wider.

The width of the Blackman window with the coefficients of Eq. (2.26) is $2L \approx 0.817T$, at half the maximum magnitude. The sidelobe of the Blackman window, as seen in Figure 2.6, is much smaller than that of the cosine-function windows with two terms, either the Hanning window or the Hamming window.

The Blackman–Harris window is an example of a cosine-function window with four terms, defined by the following coefficients (Harris, 1978; Albrecht, 2001):

$$(c_0, c_1, c_2, c_3) = (0.36358, 0.48918, 0.13660, 0.01064), \tag{2.27}$$

where $\sum_{k=0}^{K-1} c_k = 1$. The width of the Blackman–Harris window with four terms is $2L \approx 0.705T$, at half the maximum magnitude. This cosine-function window with four terms, as shown in Figure 2.7, is slightly narrower than the window with three terms. Both the Blackman window with three terms and the Blackman–Harris window with four terms are very close to the Gaussian function shown in Figure 2.3.

Figure 2.8 compares Blackman–Harris windows with high numbers of terms. The three examples are $K = \{5, 7, 9\}$, and the coefficients were given by Albrecht (2001). The width of the window becomes narrower as the number of terms increases, and accordingly the width of the frequency spectrum becomes somewhat wider. The main difference between the different windows is that as the number of cosine terms increases, the spectral leakage is reduced. This is explained in the next section.

2.4 Spectral Leakage of Window Functions

Spectral leakage is caused by the definition of the window function, and it is not caused by any discretisation of either the original seismic trace or the continuous window function.

All window functions $w(t)$ that have been discussed are symmetrical functions of time. Because of symmetry, they are zero-phased functions, and the imaginary part of their frequency spectra $\hat{w}(f)$ is zero. Therefore, the amplitude spectra $|\hat{w}(f)|$ of these window functions are the absolute values of the real part of their frequency spectra. To observe the spectral leakage behaviour of various window functions, the amplitude spectra are normalised and plotted in dB.

Figure 2.9 compares the amplitude spectra of the following four window functions: (a) the boxcar window function defined by $T = 0.25$ s, (b) the tapered boxcar window defined by $(T, \sigma) = (0.2\,\text{s}, 0.05\,\text{s})$, (c) the Gaussian function with an infinite time duration defined by the spread parameter $\sigma = 0.125$ s, and (d) the truncated Gaussian function defined by $T = 3.26\sigma$ and $\sigma = 0.125$ s.

For the boxcar window function (Figure 2.9a), the largest sidelobe of the spectrum is −13.26 dB, at frequency of ±2.86 Hz. For the tapered boxcar window function (Figure 2.9b), the largest sidelobe is −14.64 dB at frequency of ±2.69 Hz. The largest sidelobes of these two boxcar windows with and without tapering have almost the same magnitude, although the magnitude of the sidelobes decays faster when the boxcar window is tapered. However, the largest sidelobe of the truncated Gaussian function (Figure 2.9d) has been reduced significantly down to −63.99 dB, at frequency of ±6.15 Hz.

Figure 2.10 shows a comparison of the amplitude spectra of the different versions of the weighted cosine-function window. For the two-term cosine-function windows, the largest sidelobe of the Hanning window function is −31.47 dB, at frequency ±2.95 Hz (Figure 2.10a). The largest sidelobe of the Hamming window function is reduced to −43.19 dB, at frequency ±2.78 Hz (Figure 2.10b).

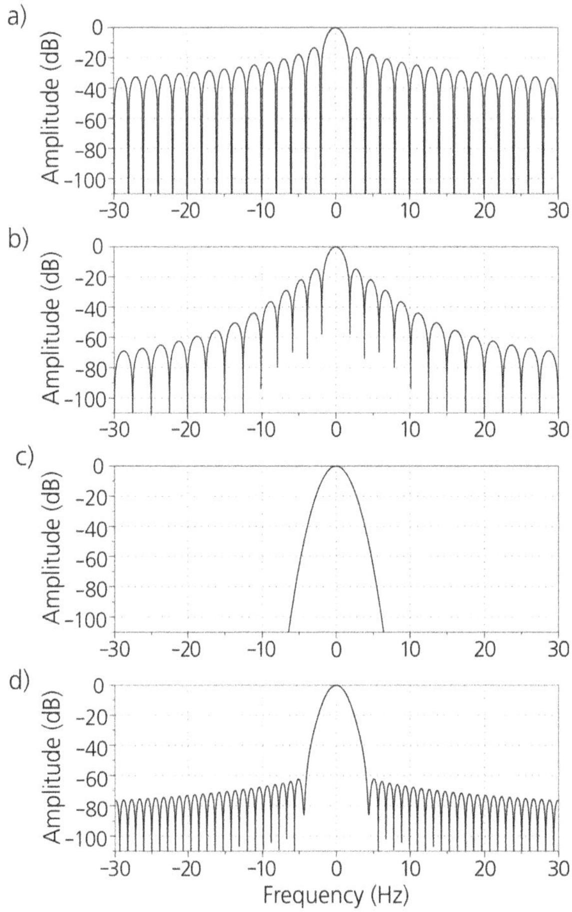

Figure 2.9 The amplitude spectra (in dB) of four window functions: (a) the boxcar window function defined by $T = 0.25$ s; (b) the tapered boxcar window function, defined by $(T, \sigma) = (0.2\ \text{s}, 0.05\ \text{s})$; (c) the Gaussian window function defined with the spread parameter $\sigma = 0.125$ s; (d) the truncated Gaussian window function, truncated at $T = 3.26\sigma$.

For the Blackman window function with three terms (Figure 2.10c), the largest sidelobe is −71.47 dB, at frequency ±3.92 Hz. For the Blackman–Harris window function with four terms (Figure 2.10d), the largest sidelobe is −97.92 dB, at frequency ±8.09 Hz. These observations on the spectral leakage are consistent with the work of Nuttall (1981).

If five cosine terms are used in a window function, the largest sidelobe can be reduced to −124.50 dB. However, if even more terms are used in the

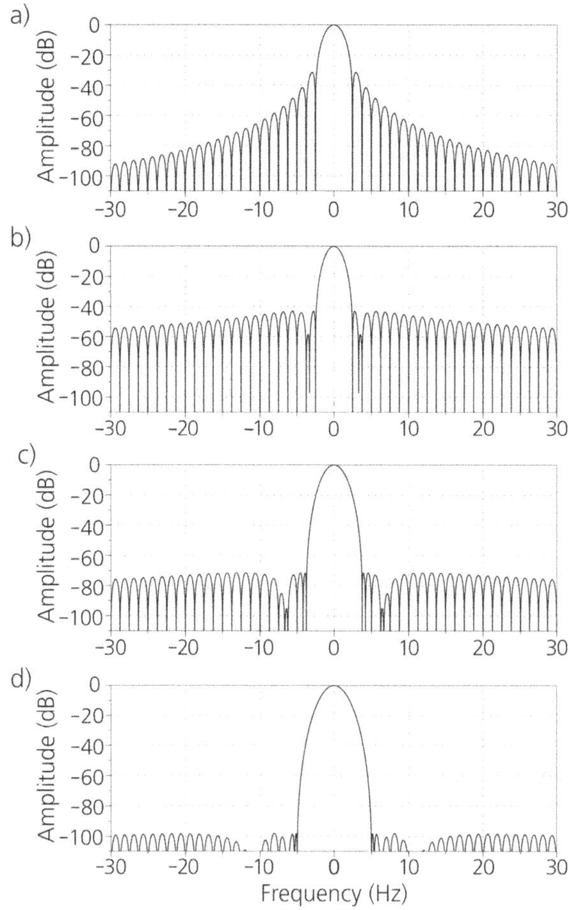

Figure 2.10 The amplitude spectra (in dB) of four cosine-function windows: (a) the Hanning window function with two terms; (b) the Hamming window function with two terms; (c) the Blackman window function with three terms; (d) the Blackman–Harris window function with four terms.

window function, the largest sidelobe is almost the same. For example, the largest sidelobes are −139.72 dB, −139.60 dB, −139.51 dB, −139.45 dB, for $K = \{6, 7, 8, 9\}$, respectively.

2.5 The Gabor Limit of Time–Frequency Resolution

In the time–frequency spectrum, there is a trade-off between the time resolution and the frequency resolution. The Gabor limit defines the maximum possible resolution as (Gabor, 1946a)

$$\sigma_t \sigma_f \geq \frac{1}{4\pi}, \tag{2.28}$$

where σ_t is the standard deviation of the time estimate (in second), and σ_f is the standard deviation of the frequency estimate (in Hz). We can prove the Gabor limit, as follows.

The standard deviation σ_f can be estimated by

$$\sigma_f = \left(\frac{\int_{-\infty}^{\infty} (f - f_m)^2 |\hat{x}(f)|^2 \, \mathrm{d}f}{\int_{-\infty}^{\infty} |\hat{x}(f)|^2 \, \mathrm{d}f} \right)^{1/2}, \tag{2.29}$$

where f_m is the mean frequency, which is the centroid (gravitational centre) of the frequency spectrum. In statistics terms, this is the second moment of the power spectrum $|\hat{x}(f)|^2$ about the mean frequency f_m. In an identical way, the standard deviation σ_t can be estimated by

$$\sigma_t = \left(\frac{\int_{-\infty}^{\infty} (t - t_m)^2 |x(t)|^2 \, \mathrm{d}t}{\int_{-\infty}^{\infty} |x(t)|^2 \, \mathrm{d}t} \right)^{1/2}, \tag{2.30}$$

where t_m is the centroid of the signal $x(t)$. This standard deviation in the time domain is quite useful in practice. For example, it can be used to quantitatively represent the short time period of a seismic wavelet (Berkhout, 1984).

Note that a standard deviation does not depend on the mean, because it defines the broadness about the mean. Therefore, we can set $f_m = t_m = 0$ in the derivation here. Although we take a signal that has zero-mean time and zero-mean frequency, there is no loss of generality. In addition, we assume that the signal $x(t)$ is normalised, $\int_{-\infty}^{\infty} |x(t)|^2 \, \mathrm{d}t = \int_{-\infty}^{\infty} |\hat{x}(f)|^2 \, \mathrm{d}f = 1$. Then, the standard deviation σ_f can be expressed as

$$\sigma_f = \left(\int_{-\infty}^{\infty} f^2 |\hat{x}(f)|^2 \, \mathrm{d}f \right)^{1/2} = \frac{1}{2\pi} \left(\int_{-\infty}^{\infty} |x'(t)|^2 \, \mathrm{d}t \right)^{1/2}, \tag{2.31}$$

where $x'(t) = \mathrm{d}x(t)/\mathrm{d}t$. The standard deviation σ_t becomes

$$\sigma_t = \left(\int_{-\infty}^{\infty} t^2 |x(t)|^2 \, \mathrm{d}t \right)^{1/2}. \tag{2.32}$$

Therefore, we have the product of two standard deviations as

$$\sigma_t \sigma_f = \frac{1}{2\pi}\left(\int_{-\infty}^{\infty} |tx(t)|^2 \,\mathrm{d}t\right)^{1/2}\left(\int_{-\infty}^{\infty} |x'(t)|^2 \,\mathrm{d}t\right)^{1/2}. \tag{2.33}$$

Now, let us consider two continuous functions $u(t)$ and $v(t)$. The Cauchy–Schwarz inequality (Schwarz, 1888) states that

$$\left(\int_{-\infty}^{\infty} |u(t)|^2 \,\mathrm{d}t\right)^{1/2}\left(\int_{-\infty}^{\infty} |v(t)|^2 \,\mathrm{d}t\right)^{1/2} \geq \left|\int_{-\infty}^{\infty} u^*(t)v(t)\mathrm{d}t\right|. \tag{2.34}$$

This inequality can be understood trivially using the definition of an inner product of two vectors, $\|\mathbf{u}\|_2 \|\mathbf{v}\|_2 \cos\alpha = \mathbf{u}\cdot\mathbf{v}$, where α is the angle between vectors $\mathbf{u}$ and $\mathbf{v}$. Because $|\cos\alpha| \leq 1$, we have $\|\mathbf{u}\|_2 \|\mathbf{v}\|_2 \geq |\mathbf{u}\cdot\mathbf{v}|$. This is the vector form of the Cauchy–Schwarz inequality.

Taking $u(t) = tx(t)$ and $v(t) = x'(t)$, Eq. (2.33) can be rewritten as

$$\sigma_t \sigma_f \geq \frac{1}{2\pi}\left|\int_{-\infty}^{\infty} tx^*(t)x'(t)\mathrm{d}t\right|. \tag{2.35}$$

Defining $x(t) = a(t)\mathrm{e}^{\mathrm{i}\theta(t)}$, the integral can be written in terms of amplitude and phase as

$$\begin{aligned}
\int_{-\infty}^{\infty} tx^*(t)x'(t)\,\mathrm{d}t &= \int_{-\infty}^{\infty} ta(t)\mathrm{e}^{-\mathrm{i}\theta(t)}\left[a'(t)\mathrm{e}^{\mathrm{i}\theta(t)} + \mathrm{i}a(t)\theta'(t)\mathrm{e}^{\mathrm{i}\theta(t)}\right]\mathrm{d}t \\
&= \int_{-\infty}^{\infty}\left[ta(t)a'(t) + \mathrm{i}ta^2(t)\theta'(t)\right]\mathrm{d}t \\
&= \frac{1}{2}\int_{-\infty}^{\infty}\left[\frac{\mathrm{d}}{\mathrm{d}t}[ta^2(t)]\right]\mathrm{d}t - \frac{1}{2}\int_{-\infty}^{\infty} a^2(t)\mathrm{d}t + \mathrm{i}\int_{-\infty}^{\infty}\left[ta^2(t)\theta'(t)\right]\mathrm{d}t.
\end{aligned} \tag{2.36}$$

There are three integrals. The first integral term is an integral of an exact differential and, thus, it integrates to zero:

$$\int_{-\infty}^{\infty} \frac{\mathrm{d}}{\mathrm{d}t}[ta^2(t)]\,\mathrm{d}t = 0. \tag{2.37}$$

The second integral term gives a unity, since the signal $x(t)$ is normalised and $\int_{-\infty}^{\infty} |x(t)|^2 \,\mathrm{d}t = \int_{-\infty}^{\infty} a^2(t)\mathrm{d}t = 1$.

In the third integral term, $\theta'(t) = 2\pi f_{\text{inst}}(t)$, where $f_{\text{inst}}(t)$ is the instantaneous frequency (Eq. 1.34). Therefore, the third integral gives the covariance of the complex signal $x(t) = a(t)e^{i\theta(t)}$ as

$$\int_{-\infty}^{\infty} ta^2(t)\theta'(t)\mathrm{d}t = 2\pi \int_{-\infty}^{\infty} t|x(t)|^2 f_{\text{inst}}(t)\mathrm{d}t = 2\pi \mathrm{cov}_{t,f}. \tag{2.38}$$

A normalisation of the covariance $\mathrm{cov}_{t,f}$ is the correlation coefficient,

$$\phi_{t,f} = \frac{\mathrm{cov}_{t,f}}{\sigma_t \sigma_f}, \tag{2.39}$$

which is a good indication of the relationship between the time t and the instantaneous frequency $f_{\text{inst}}(t)$.

Hence, we can prove Eq. (2.28) by rewriting Eq. (2.35) as the following:

$$\sigma_t \sigma_f \geq \frac{1}{2\pi}\left| \int_{-\infty}^{\infty} tx^*(t)x'(t)\mathrm{d}t \right| = \left| -\frac{1}{4\pi} + \mathrm{i}\,\mathrm{cov}_{t,f} \right| = \left(\frac{1}{(4\pi)^2} + \mathrm{cov}_{t,f}^2 \right)^{1/2} \geq \frac{1}{4\pi}. \tag{2.40}$$

The Gabor limit of Eq. (2.28) states that the product of time resolution and frequency resolution is at least $(4\pi)^{-1}$, which is 80 millicycles (or ms·Hz). This Gabor limit is analogous to Heisenberg's uncertainty principle, originally expressed in terms of observation and measurement (Heisenberg, 1927), and distinguished by a scale for the quantum dimensions, called Planck's constant. However, the physical meaning of the Gabor limit in signal analysis is different. The energy density of the signal in time is $|x(t)|^2$, and the energy density in frequency is $|\hat{x}(f)|^2$. The Gabor limit states that if one density is narrow, the other is wide. The point is not that both time and frequency (as measurements) cannot be made arbitrarily narrow, but that time density and frequency density cannot be made narrow at the same time (Hall, 2006).

2.6 Implementation of the Gabor Transform

The physical meaning of the Gabor transform can be understood as a time convolution process, at any given frequency f, between a complex signal $y(t, f) = x(t)e^{-i2\pi ft}$ and a window function $w(t)$:

$$S_{\text{GT}}(t, f) = x(t)e^{-i2\pi ft} * w(t) = y(t, f) * w(t), \tag{2.41}$$

where $*$ is the convolution operator with respect to time t. The Fourier transform of the complex signal $y(t, f)$ is

$$\hat{y}(\xi, f) = \mathcal{F}_t\left[y(t, f)\right] = \mathcal{F}_t\left[x(t)e^{-i2\pi ft}\right], \tag{2.42}$$

where $\mathcal{F}_t$ is the Fourier transform with respect to time t, and ξ is the corresponding frequency variable. A fast Fourier transform (FFT) can be used in the implementation but note that the FFT here is a complex-to-complex transform.

Previous sections have derived analytically the Fourier transforms of various window functions. Specifically, Eqs. (2.5), (2.8), (2.18), and (2.23) are the Fourier transforms of the boxcar window, the tapered boxcar window, the truncated Gaussian window, and the multiple cosine-term window, respectively.

Then, the time-convolution form of the Gabor transform in Eq. (2.41) can be implemented through forward and inverse Fourier transforms:

$$S_{GT}(t, f) = \mathcal{F}_\xi^{-1}\left[\hat{y}(\xi, f)\hat{w}(\xi)\right], \tag{2.43}$$

where $\mathcal{F}_\xi^{-1}$ is the inverse Fourier transform with respect to frequency ξ. Note again that $\mathcal{F}_\xi^{-1}$ is a complex-to-complex inverse transform.

Let us compare the time–frequency spectra that are generated by the Gabor transform using different window functions. We apply the Gabor transform to the same seismic trace, which is taken between 1 s and 2 s, with the time sampling interval of 2 ms. We compare the spectra in four groups, to demonstrate the effectiveness of various window functions.

The first group, shown in Figure 2.11, consists of the time–frequency spectra produced by the Gabor transform with a boxcar window function (without tapering) and a tapered boxcar window function, respectively. In the boxcar window function, the (half-)length of the time window is $T = 50$ ms. In the tapered boxcar window function, the (half-)lengths of the boxcar and the taper are 40 ms and 10 ms, respectively. Thus, the (half-) length of the time window is also 50 ms. The time–frequency spectra using both window functions have strong sidelobes. The highest sidelobes are approximately −13 dB and −14 dB, respectively (as shown in Figures 2.9a and 2.9b).

The second group, shown in Figure 2.12, consists of the time–frequency spectra that are generated by the Gabor transform with the Hanning and

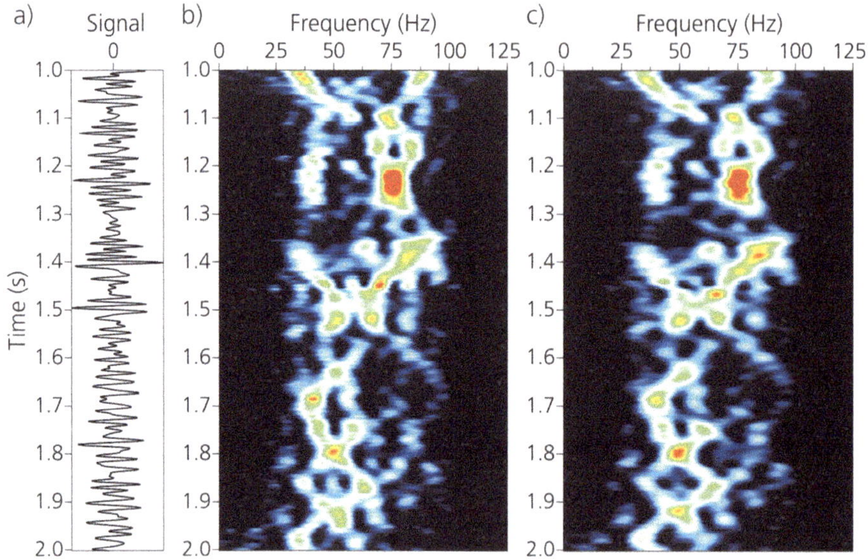

Figure 2.11 A field seismic trace (a) and the corresponding time–frequency spectra generated with the Gabor transform, using the boxcar window function (b) and the tapered boxcar window function (c), respectively.

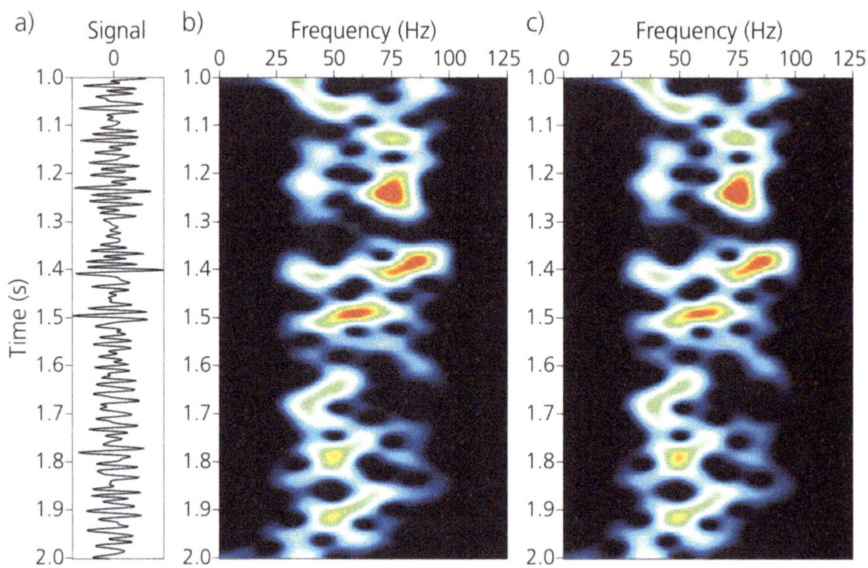

Figure 2.12 The field seismic trace (a) and the corresponding time–frequency spectra generated with the Gabor transform, using the Hanning window function (b) and the Hamming window function (c), respectively.

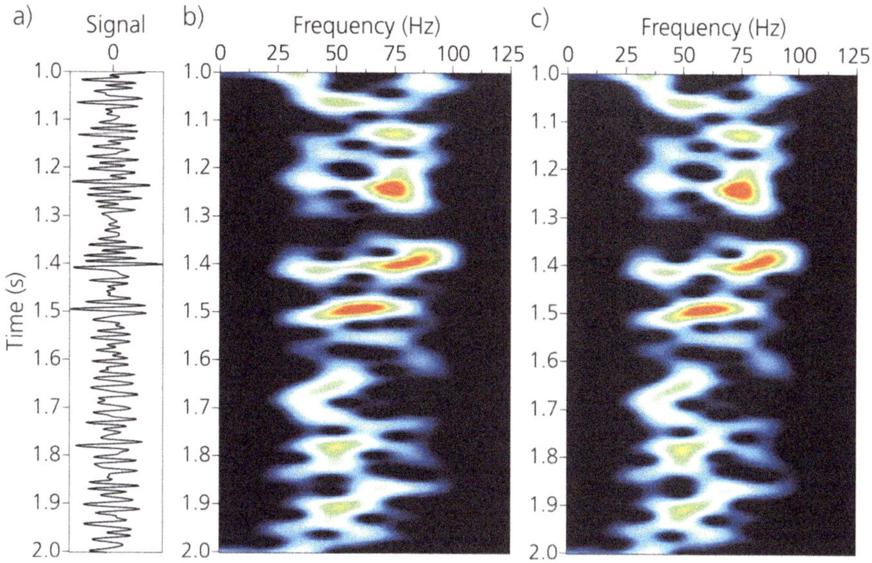

Figure 2.13 The field seismic trace (a) and the corresponding time–frequency spectra generated by the Gabor transform, using the truncated Gaussian function (b) and the Blackman window with three terms (c), respectively.

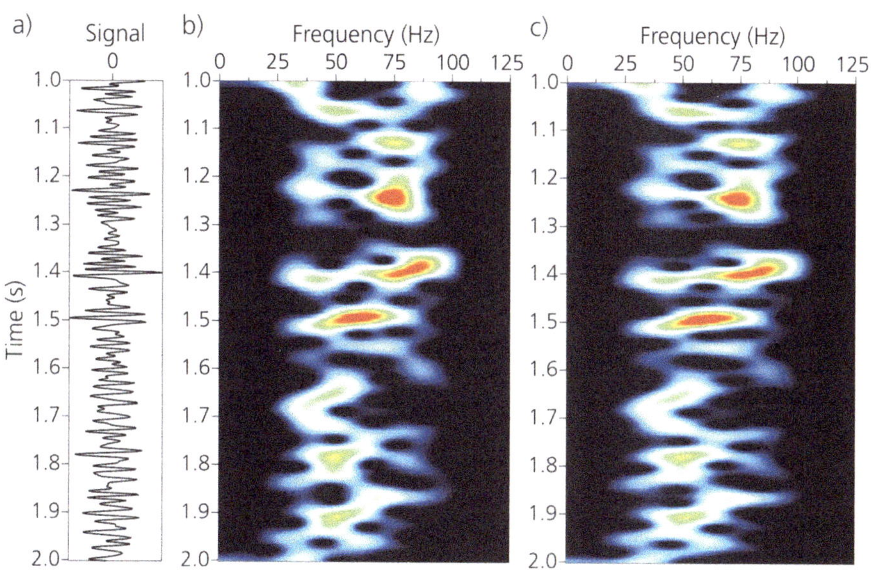

Figure 2.14 The field seismic trace (a) and the corresponding time–frequency spectra generated by the Gabor transform, using the Blackman window with three terms (b) and the Blackman–Harris window with four terms (c), respectively.

Hamming windows. These two window functions are composed of two terms. The spectral characteristics, including the largest sidelobes, of the Hanning and Hamming windows are very similar (as shown in Figures 2.10a and 2.10b). The time–frequency spectra are therefore similar.

The third group, shown in Figure 2.13, includes the time–frequency spectra produced by the Gabor transform with the truncated Gaussian function and the Blackman window function. According to Eq. (2.15), the Gaussian spread parameter σ is related to the window length T through the relationship,

$$\sigma = \frac{1}{3.26} T. \tag{2.44}$$

As we can see from the frequency spectra (Figures 2.9d and 2.10c), the truncated Gaussian function and the Blackman window function are very similar. Therefore, the time–frequency spectra produced by the Gabor transform with these two windows are also similar.

The fourth group, shown in Figure 2.14, includes the time–frequency spectra produced by the Gabor transform with the Blackman window function with three terms and the Blackman–Harris window function with four terms. The frequency spectrum of the Blackman–Harris window function with four terms (Figure 2.10d) has a slightly wider bandwidth, and a smaller sidelobe, than the spectrum of the Blackman window function with three terms (Figure 2.10c). Therefore, the time–frequency spectrum generated by the Gabor transform, using the Blackman–Harris window function with four terms (Figure 2.14c), shows a slightly smoother pattern along the frequency axis than the spectrum using the Blackman window function with three terms (Figure 2.14b).

2.7 The Inverse Gabor Transform

After the data processing in the time–frequency spectrum produced by the Gabor transform, we may need to reconstruct a seismic trace in the time domain from the processed time–frequency spectrum. This is the inverse Gabor transform.

In the following, we will discuss a two-step implementation of the inverse Gabor transform. The first step is to perform the inverse Fourier transform to recover the windowed trace segment $s(t, \tau)$ at any given time t (Allen & Rabiner, 1977), as

$$s(t,\tau)=\int_{-\infty}^{\infty} S_{\mathrm{GT}}(t,f)\mathrm{e}^{\mathrm{i}2\pi f\tau}\mathrm{d}f. \tag{2.45}$$

The second step is to reconstruct the seismic trace from $s(t,\tau)$ by integration over time t:

$$x(\tau)=h(\tau)\int_{-T}^{T} s(t,\tau)\mathrm{d}t, \tag{2.46}$$

where $h(\tau)$ is a synthesis scaling factor, and $x(\tau)$ is equivalent to $x(t)$.

The synthesis scaling factor $h(\tau)$ can be determined in terms of the given analysis window $w(t)$. Because $s(t,\tau)=w(t-\tau)x(\tau)$, Eq. (2.46) can be written as

$$x(\tau)=h(\tau)\int_{-T}^{T} w(t-\tau)x(\tau)\mathrm{d}t=x(\tau)\left(h(\tau)\int_{-T}^{T} w(t-\tau)\mathrm{d}t\right). \tag{2.47}$$

Since $w(t-\tau)$ is stationary with respect to time τ, Eq. (2.47) leads to an invertibility condition as

$$h(\tau)=\left(\int_{-T}^{T} w(t,\tau)\mathrm{d}t\right)^{-1}. \tag{2.48}$$

Figure 2.15 illustrates the derivation of the synthesis scaling factor $h(\tau)$ at the time position τ. In the first step, the signal segments $s(t,\tau)$ are recovered (the dotted vertical lines) by the inverse Fourier transform $S(t,f)\rightarrow s(t,\tau)$. The analysis window function $w(t-\tau)$ that was applied to the original seismic trace is plotted as a red curve.

The second step is an integral of $w(t-\tau)$ over time t, at any given time position τ. The weighting coefficients consecutively that are multiplied to the original signal (the vertical dashed line) are given by $w(t,\tau)$. The synthesis scaling factor $h(\tau)$ in the inverse Gabor transform is for removing the weighting coefficients $w(t,\tau)$ at time τ. Thus, the synthesis scaling factor $h(\tau)$ can be built in terms of $w(t,\tau)$, as presented in Eq. (2.48).

Practically, the invertibility condition in the form of Eq. (2.48) is useful in the discrete-time domain, for reducing any potential errors resulting from the discretisation and truncation of the analysis window function. The sampling of time t should meet the Nyquist criterion, so that the analysis window function $w(t,\tau)$ can be appropriately integrated over time t for constructing a synthesis scaling factor $h(\tau)$ (Zhivomirov, 2019).

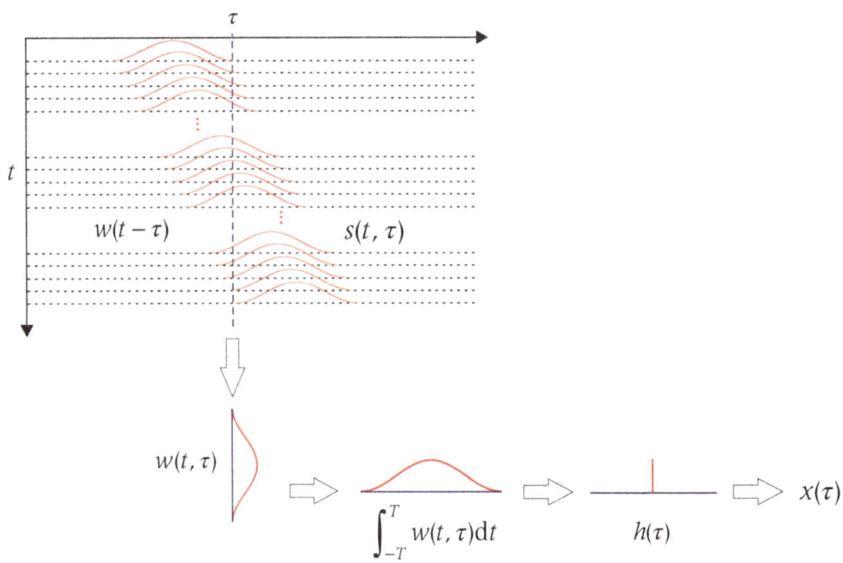

Figure 2.15 The synthesis scaling factor for the inverse Gabor transform. Each horizontal dotted line at time t represents a signal segment $s(t,\tau)$, obtained by an inverse Fourier transform $S(t,f) \rightarrow s(t,\tau)$. Each red curve that varies along the horizontal axis τ is an analysis window $w(t-\tau)$ that was successively applied to the original signal $x(t)$, to form the segment $s(t,\tau)$. At a certain time position τ, the weighting coefficients of the applied analysis windows along the vertical dashed line form a window $w(t,\tau)$, and its integral gives the synthesis scaling factor $h(\tau)$.

In addition, the trace reconstruction with a smaller error by the inverse Gabor transform requires a high time–frequency spectrum concentration as well. That is, the window function $w(t)$ should be wide enough so that the resultant Gabor spectrum $S_{GT}(t,f)$ has an adequate spectral localisation over the frequency axis (Zielinski, 2001).

2.8 Application in Inverse *Q* Filtering

This section will demonstrate an application of the Gabor transform in seismic inverse Q filtering (Wang, 2006, 2008).

The procedure of the Gabor transform-based inverse Q filtering consists of the following five steps:

(1) For a given seismic trace $x(t)$, constructing a series of the signal segments $s(t,\tau)$.

(2) Performing the Fourier transforms on each segment $s(t,\tau)$ with respect to time τ, generating the Gabor transform spectrum $S(t,f)$, which represents the input to the inverse Q filtering process.
(3) Implementing inverse Q filtering on $S(t,f)$ to produce a processed time–frequency spectrum $\tilde{S}(t,f)$.
(4) Performing the inverse Fourier transforms on $\tilde{S}(t,f)$ with respect to frequency f, to reconstruct signal segments $\tilde{s}(t,\tau)$.
(5) Integrating over time t on the set of the signal segments $\tilde{s}(t,\tau)$, to reproduce the processed seismic trace $\tilde{x}(\tau)$ in the time domain.

Steps 1 and 2 are the forward Gabor transform, and Steps 4 and 5 are the inverse Gabor transform. Step 3 is the core of the inverse Q filtering algorithm, which can be summarised explicitly as follows.

Given the Gabor transform spectrum $S(t,f)$ at a given time t, the inverse Q filter can be expressed in the time–frequency domain as

$$\tilde{S}(t,f) = S(t,f)\Lambda(t,f)\mathrm{e}^{\mathrm{i}\int_0^t 2\pi f\left[(f/f_h)^{-\gamma_Q(\tau)}-1\right]\mathrm{d}\tau}, \tag{2.49}$$

where $\Lambda(t,f)$ is the operator for amplitude compensation, the exponential function is the operator for phase correction, f_h is a reference frequency, $\gamma_Q(t) = [\pi Q(t)]^{-1}$, and $Q(t)$ is the quality factor of the subsurface anelastic media.

Equation (2.49) is a stable inverse Q filter. The stabilised operator for amplitude compensation is given by

$$\Lambda(t,f) = \frac{A(t,f)+\mu^2}{A^2(t,f)+\mu^2}, \tag{2.50}$$

where μ^2 is the stabilisation factor, and $A(t,f)$ is the attenuation factor,

$$A(t,f) = \mathrm{e}^{-\int_0^t \frac{\pi f}{Q(\tau)}\left(\frac{f}{f_h}\right)^{-\gamma_Q(\tau)}\mathrm{d}\tau}. \tag{2.51}$$

Figure 2.16a is a synthetic seismic trace with $Q = 88$ and the Gabor transform spectrum, generated using the Blackman–Harris window function that is defined by the four-term cosine functions. Because of the Q^{-1} effect, the amplitude and phase of the wavelets change with the travel time. After inverse Q filtering (Eq. 2.49), the time–frequency spectrum is modified. Figure 2.16b shows that all wavelets have been restored to the zero phase, and the bandwidths at various times have been expanded.

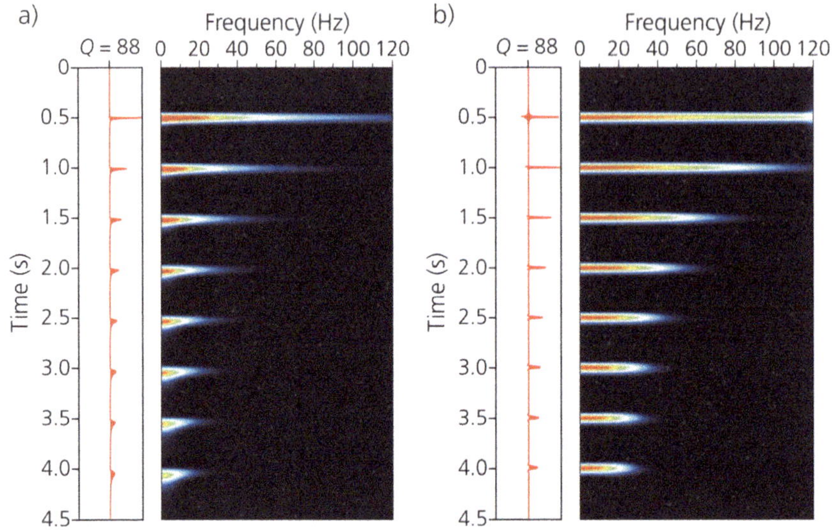

Figure 2.16 The inverse Q filtering based on the Gabor transform. (a) A synthetic trace with $Q = 88$, clearly showing that the amplitude and phase of the wavelets change with time, and the corresponding time–frequency spectrum. (b) The seismic trace after inverse Q filtering, and the corresponding time–frequency spectrum.

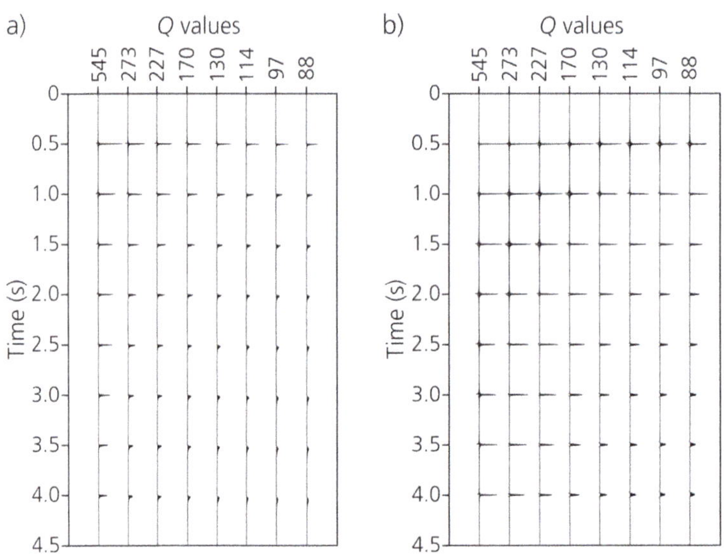

Figure 2.17 The inverse Q filtering based on the Gabor transform. (a) Synthetic seismic traces with different Q values. The amplitude and phase of the wavelets change with the Q value and travel time. (b) The seismic traces after inverse Q filtering. The amplitudes of the wavelets are partially recovered, but all wavelets appear as zero phase.

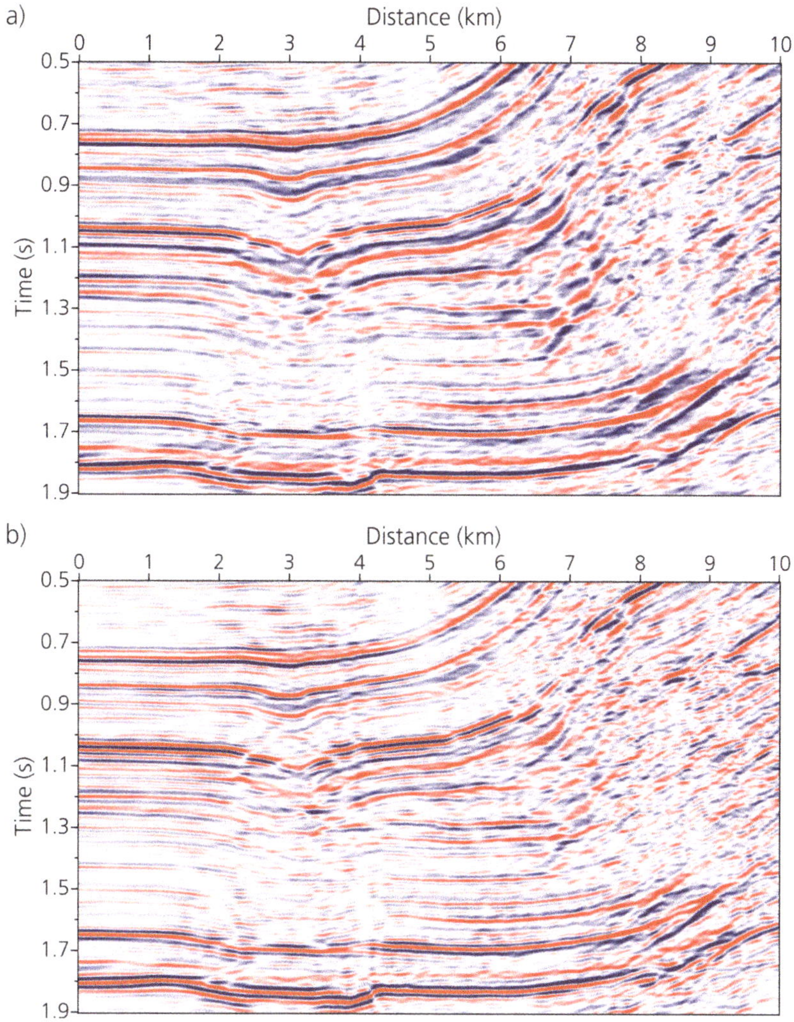

Figure 2.18 The effectiveness of inverse Q filtering. (a) An example of a field seismic profile. (b) The seismic profile after inverse Q filtering, which compensates for amplitude attenuation and corrects for phase distortion.

However, the amplitude spectra of most wavelets could be partially restored. This is because the stabilisation in the inverse Q filter automatically restricts the attempt to recover the high-frequency components when these components were completely attenuated in the original seismic signals (Wang, 2002). The stabilisation strategy avoids the amplification of ambient noise and machine errors, which are associated with numerical instability.

Figure 2.17 is a comparison of the synthetic traces before and after inverse Q filtering. The synthetic traces show that the phases and amplitudes of the wavelets change with the Q value and the travel time. After inverse Q filtering based on the Gabor transform, the amplitudes of the frequency components were stably restored, and the phases of all frequency components were corrected. All wavelets appear as zero phase.

Figure 2.18 illustrates the effectiveness of inverse Q filtering on a field seismic profile. After inverse Q filtering based on the Gabor transform, the seismic profile looks as follows:

(1) Vertical resolution has been improved. All waveforms were narrowed, and thin-layered reflections with weak amplitudes were separated. The separation could reflect the different thickness of the target layers.
(2) Lateral continuation was enhanced, and spatial variation of amplitudes became visible. Due to the undisturbed sedimentary environment, thin-layered reflections can be clearly traced across the seismic profile with good continuity. The spatial variation of amplitude may implicitly reflect the distribution of the thin-layered reservoir properties.
(3) Faults and spatial discontinuities have been identified. Details of small faults and fractures will facilitate geological interpretation.
(4) Travel time has also been corrected. Inverse Q filtering not only compensates for amplitude attenuation, but also corrects for phase distortion. The phase correction appears as a continuous time shift along the time axis. Therefore, the two-way travel time better matches the true geological depth.

3

The Continuous Wavelet Transform

In the Gabor transform with various window functions, including the short-time Fourier transform (STFT) with a boxcar window, one shortcoming is that the window size is constant. The continuous wavelet transform differs significantly from the Gabor transform, because it uses a windowing technique with variable window sizes for different frequencies. Consequently, it allows a clear localisation of the frequency components present in the nonstationary signal on the time axis.

The wavelet used in the continuous wavelet transform has a dilation parameter to continuously dilate the wavelet. This is where the name 'continuous' comes from. Therefore, the continuous wavelet transform provides an overcomplete representation of the nonstationary seismic signal in the time–frequency plane.

3.1 Basics of the Continuous Wavelet Transform

The history of the development of the continuous wavelet transform goes back to Gabor's concept of the 'elementary signal', defined as follows (Gabor, 1946a; Helstrom, 1966; Bastiaans, 1980)

$$\psi(t,f) = \frac{1}{\sqrt[4]{\pi\sigma^2}} e^{-\frac{t^2}{2\sigma^2}} e^{i2\pi ft}. \tag{3.1}$$

Gabor's elementary signal defines a Gaussian envelope used to constrain a complex harmonic function of frequency f. This complex signal is called a Gabor wavelet (Figure 3.1a).

Time–Frequency Analysis of Seismic Signals, First Edition. Yanghua Wang.

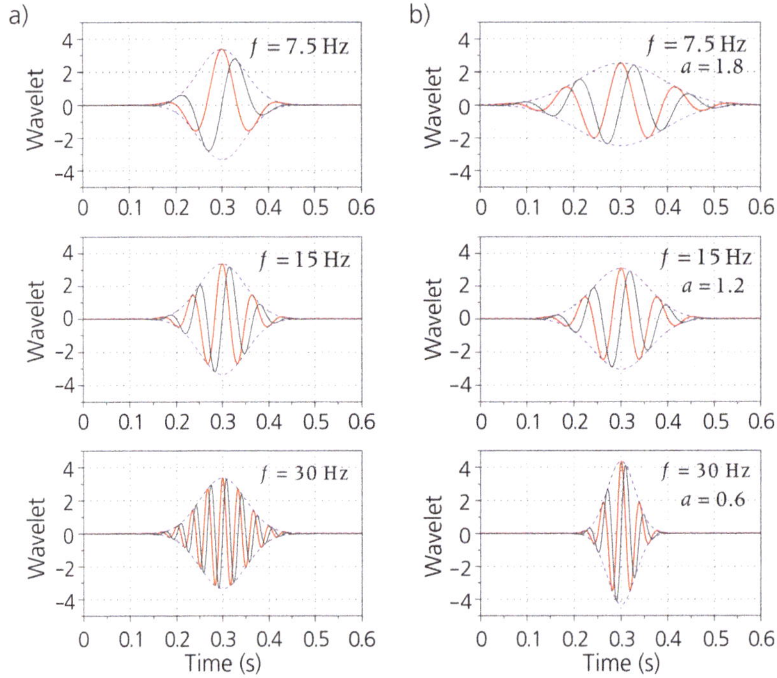

Figure 3.1 The Gabor wavelets and the dilated wavelets. (a) The Gabor wavelets, in which a Gaussian envelope constrains a complex harmonic function. The real and imaginary parts are shown in red and black curves, respectively. (b) The dilation effect. When the dilation parameter $a > 1$, the dilated wavelet is stretched. When the dilation parameter $a < 1$, the dilated wavelet is contracted more.

At any given time t, the Gabor expansion (GE) of seismic signal $x(t)$ is expressed in integral form as

$$S_{\mathrm{GE}}(t, f) = \int_{-\infty}^{\infty} x(\tau)\psi^*(\tau - t, f)\mathrm{d}\tau, \tag{3.2}$$

where $\psi^*(t, f)$ is the complex conjugate of $\psi(t, f)$. Substituting the Gabor wavelet of Eq. (3.1) into Eq. (3.2), we may obtain the formula:

$$S_{\mathrm{GE}}(t, f) = \frac{1}{\sqrt[4]{\pi\sigma^2}} \int_{-\infty}^{\infty} x(\tau)\mathrm{e}^{-\frac{(\tau-t)^2}{2\sigma^2}} \mathrm{e}^{-\mathrm{i}2\pi f(\tau-t)}\mathrm{d}\tau. \tag{3.3}$$

Comparing this Gabor expansion to the Gabor transform (GT) defined in Eq. (2.13), we find that

$$S_{GE}(t,f) = e^{i2\pi ft} S_{GT}(t,f). \tag{3.4}$$

The Gabor expansion $S_{GE}(t,f)$ and the Gabor transform $S_{GT}(t,f)$ differ in the phase-shift term $e^{i2\pi ft}$. Substituting $u = \tau - t$, the Gabor expansion in Eq. (3.3) can also be written as

$$S_{GE}(t,f) = \frac{1}{\sqrt[4]{\pi\sigma^2}} \int_{-\infty}^{\infty} x(t+u) e^{-\frac{u^2}{2\sigma^2}} e^{-i2\pi fu} du. \tag{3.5}$$

Because of the phase-shift term, the origin of the signal is translated to the given time t, $x(t+u)|_{u=0} = x(t)$, and the Gaussian envelope is concentrated around t, prior to the implementation of Fourier transform. Therefore, the Gabor expansion $S_{GE}(t,f)$, which uses the Gabor wavelet, produces a localised frequency spectrum, and minimise the uncertainty in information.

Taking this advantage of the Gabor expansion, a group of geophysicists, Pierre Goupillaud, Alexander Grossmann, and Jean Morlet, proposed to further modify the Gabor expansion and introduce the continuous wavelet transform (Goupillaud *et al.*, 1984; Grossmann & Morlet, 1984; Selesnick *et al.*, 2005). The natural difference between the Gabor expansion and the continuous wavelet transform is that the time resolution of the Gabor expansion is fixed, but the time resolution of the continuous wavelet transform varies along the frequency axis.

For a given signal $x(t)$, modifying the Gabor expansion of Eq. (3.2), the continuous wavelet transform (CWT) can be expressed in the form

$$S_{CWT}(t,a) = \int_{-\infty}^{\infty} x(\tau) \frac{1}{\sqrt{|a|}} \psi^* \left(\frac{\tau - t}{a} \right) d\tau, \tag{3.6}$$

in which the timescale a is the dilation parameter. While the translation time t serves for localisation, which is the attractive feature of Gabor expansion, the dilation parameter a either dilates or contracts the wavelet.

The original wavelet $\psi(t)$ is called the mother wavelet, which generates daughter wavelets by dilation (Figure 3.1b). In the continuous wavelet transform only a positive dilation parameter $a > 0$ is considered here. A negative dilation parameter corresponds to a time-reversed wavelet (Olhede & Walden, 2002). For $a > 0$, the dilated wavelet is denoted as

$$\psi_a(t) = \frac{1}{\sqrt{a}} \psi \left(\frac{t}{a} \right), \tag{3.7}$$

where $\sqrt{a}$ is the normalisation factor for ensuring energy conservation of the dilated wavelet $\psi_a(t)$:

$$\int_{-\infty}^{\infty} |\psi_a(t)|^2 \mathrm{d}t = \int_{-\infty}^{\infty} |\psi(t)|^2 \mathrm{d}t. \tag{3.8}$$

The notion of the dilation parameter a is inherently related to time resolution in signal analysis. Recall that the continuous wavelet transform is a cross-correlation, that measures the similarity between the signal $x(t)$ and a dilated wavelet $\psi_a(t)$. When the dilation parameter is relatively small, $0 < a < 1$, the dilated wavelet $\psi_a(t)$ is more contracted. The cross-correlation leads to a more detailed graphical representation for the given signal. However, the signal components with a small timescale do not last for the entire duration of the signal and, on the other hand, dilated wavelets with a relatively large timescale are also required in the signal analysis. If the dilation parameter is high, $a > 1$, the dilated wavelet $\psi_a(t)$ is stretched, and the cross-correlation represents the signal in less detail.

In the continuous wavelet transform, the mother wavelet $\psi(t)$ is a complex function, or analytic wavelet that has a spectrum with only positive frequencies. Since an analytic wavelet responds only to the non-negative frequencies of a given signal, it produces a transform whose modulus is less oscillatory than that of a real wavelet. This property is an attractive advantage in detecting and tracking instantaneous frequencies of the nonstationary seismic signal.

The continuous wavelet transform can be treated as a time-domain convolution (Mallat, 2009):

$$S_{\mathrm{CWT}}(t,a) = \int_{-\infty}^{\infty} x(\tau)\psi_a^*(\tau - t)\mathrm{d}\tau = \int_{-\infty}^{\infty} x(\tau)\bar{\psi}_a(t-\tau)\mathrm{d}\tau, \tag{3.9}$$

where $\bar{\psi}_a(t)$ is a reflected version of the dilated wavelet $\psi_a(t)$,

$$\bar{\psi}_a(t) = \frac{1}{\sqrt{a}}\psi^*\left(\frac{-t}{a}\right). \tag{3.10}$$

The convolution amounts to filtering. While the time duration of filter $\bar{\psi}_a(t)$ changes with the dilation parameter a, the frequency bandwidth changes accordingly. Therefore, the continuous wavelet transform is equivalent to bandpass filtering, which uses different passband widths, and is applied to the Fourier transform $\hat{x}(f)$ of signal $x(t)$.

The Fourier transform of $\bar{\psi}_a(t)$ is

$$\hat{\bar{\psi}}_a(f) = \sqrt{a}\hat{\psi}^*(af) = \hat{\psi}_a^*(f). \tag{3.11}$$

Then, the time-domain convolution of Eq. (3.9) equals to the product of $\hat{x}(f)$ and $\hat{\psi}_a^*(f)$ in the frequency domain. The continuous wavelet transform can be implemented based on the inverse Fourier transform (Torrence & Compo, 1998; Addison *et al.*, 2009; Montejo & Suarez, 2013; Komorowski & Pietraszek, 2016):

$$S_{\mathrm{CWT}}(t,a) = \int_{-\infty}^{\infty} \hat{x}(f)\hat{\psi}_a^*(f)\mathrm{e}^{\mathrm{i}2\pi ft}\,\mathrm{d}f. \tag{3.12}$$

For the implementation of this equation, a fast Fourier transform (FFT) algorithm can be employed to speed up the computation and to save considerable computing power.

For the continuous wavelet transform, the primary condition of a mother wavelet is that its average must be zero:

$$\int_{-\infty}^{\infty} \psi(t)\mathrm{d}t = 0. \tag{3.13}$$

This zero-average condition indicates that the mother wavelet (when $a = 1$) oscillates around zero. The energy of a mother wavelet, either in the time domain or the frequency domain, is normalised as (Helstrom, 1966)

$$\int_{-\infty}^{\infty} |\psi(t)|^2\,\mathrm{d}t = \int_{-\infty}^{\infty} |\hat{\psi}(f)|^2\,\mathrm{d}f = 1. \tag{3.14}$$

This finite energy condition indicates that the amplitude of the time-domain wavelet diminishes quickly. For example, the Ricker wavelet of Eq. (1.41) satisfies these two conditions listed in Eqs. (3.13–3.14).

3.2 The Complex Morlet Wavelet

The continuous wavelet transform is a cross-correlation between the signal and a dilated wavelet. Therefore, the choice of the mother wavelet is crucial for a successful time–frequency analysis with the continuous wavelet transform (Daubechies, 1992; Mallat, 2009). Matching the nonstationary seismic signals, the continuous wavelet transform often uses an analytic wavelet, where only non-negative frequencies need to be considered.

The complex Morlet wavelet is a mother wavelet that is commonly used in seismic signal analysis. The Morlet wavelet is a Gabor wavelet, defined as a complex sinusoid modulated by a Gaussian envelope:

$$\psi_\sigma(t) = \sqrt[4]{\frac{(8\ln 2)f_c^2}{\pi\sigma^2}}\, \mathrm{e}^{-\frac{(4\ln 2)f_c^2 t^2}{\sigma^2}}\, \mathrm{e}^{\mathrm{i}2\pi f_c t}, \tag{3.15}$$

where f_c is the central frequency of the Morlet wavelet, and σ is the spread parameter controlling the (half-)breadth of the wavelet. The ratio of the central frequency f_c to the Gaussian spread parameter σ determines the number of oscillations of the complex sinusoid inside the (half-) Gaussian envelope.

A significant feature of the Morlet wavelet definition is the use of spread parameter σ, which is designed specifically for seismic signals in which the wavelet has relatively a short time duration. To assure that the definition in Eq. (3.15) is a 'wavelet', the spread parameter σ shall be set as $\sigma \geq 2$. The time integral of Eq. (3.15) can be derived as

$$\int_{-\infty}^{\infty} \psi_\sigma(t)\mathrm{d}t = \sqrt[4]{\frac{\pi\sigma^2}{(2\ln 2)f_c^2}}\, \mathrm{e}^{-\frac{\pi^2\sigma^2}{4\ln 2}}. \tag{3.16}$$

It can be verified that $\int_{-\infty}^{\infty} \psi_\sigma(t)\mathrm{d}t < 10^{-6}$ if $\sigma \geq 2$. Therefore, the Morlet wavelet fulfils the zero-average condition.

The Morlet wavelet is a complex wavelet, but its Fourier transform is simply real:

$$\hat{\psi}_\sigma(f) = \sqrt[4]{\frac{\pi\sigma^2}{(2\ln 2)f_c^2}}\, \mathrm{e}^{-\frac{\pi^2\sigma^2(f-f_c)^2}{(4\ln 2)f_c^2}}. \tag{3.17}$$

Figure 3.2 shows an example Morlet wavelet (red) defined by $f_c = 30$ Hz and $\sigma = 5$. The amplitude spectrum of the Morlet wavelet is concentrated at the central frequency (Figure 3.2a). Applying a dilation parameter $a = 2$, the central frequency is reduced from 30 Hz to 15 Hz, and the frequency bandwidth is narrowed. Correspondingly, the dilated Morlet wavelet becomes more spread out (Figure 3.2b).

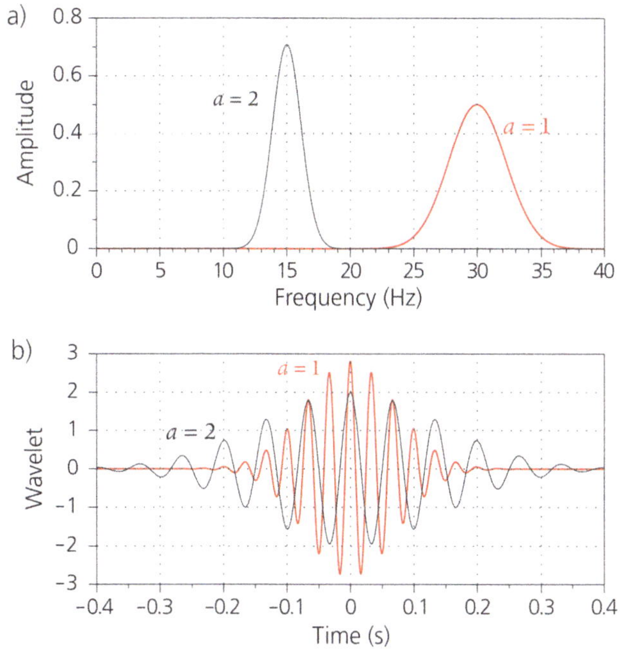

Figure 3.2 The Morlet wavelet, defined with the central frequency $f_c = 30$ Hz and the Gaussian spread parameter $\sigma = 5$. (a) The amplitude spectrum $\hat{\psi}(f)$ of the Morlet wavelet (red) and a dilated spectrum $\hat{\psi}_a(f) = \sqrt{a}\,\hat{\psi}(af)$ with $a = 2$ (black). (b) The real part of the Morlet wavelet $\psi(t)$ and the dilated wavelet $\psi_a(t)$ in the time domain.

When the amplitude spectrum of the dilated wavelet is highly concentrated at a fixed frequency value, the continuous wavelet transform will have the tendency to concentrate at the frequency values associated with the dominant harmonics of the signal, defining a series of ridges that evolve over time. These ridges are locations at which the central frequency of a dilated wavelet coincides with the local frequency of the nonstationary signal.

According to the frequency-domain formulation of Eq. (3.12), the continuous wavelet transform can be implemented by an inverse Fourier transform:

$$S_{\mathrm{CWT}}(t,a) = \sqrt[4]{\frac{\pi\sigma^2 a^2}{(2\ln 2)f_c^2}} \int_{-\infty}^{\infty} \hat{x}(f) e^{-\frac{\pi^2\sigma^2(af-f_c)^2}{(4\ln 2)f_c^2}} e^{i2\pi ft} df. \qquad (3.18)$$

3.3 The Complex Morse Wavelet

The Morse wavelet is another mother wavelet that is widely used in the continuous wavelet transform and has good localisation properties (Olhede & Walden, 2002).

The standard zeroth order Morse wavelet is defined in the frequency domain as

$$\hat{\psi}^{(0)}_{\beta,\gamma}(f) = \sqrt{\frac{2^{\alpha+1}\pi\gamma}{\Gamma(\alpha)}}(2\pi f)^{\beta}\,\mathrm{e}^{-(2\pi f)^{\gamma}}, \qquad \text{for } f \geq 0, \tag{3.19}$$

where β is the frequency-domain compactness parameter, γ describes the frequency-domain asymmetry of the Morse wavelet, and $\alpha = (1+2\beta)/\gamma$. The superscript '(0)' of $\hat{\psi}^{(0)}(f)$ indicates the order of the Morse wavelet (Daubechies & Paul, 1988).

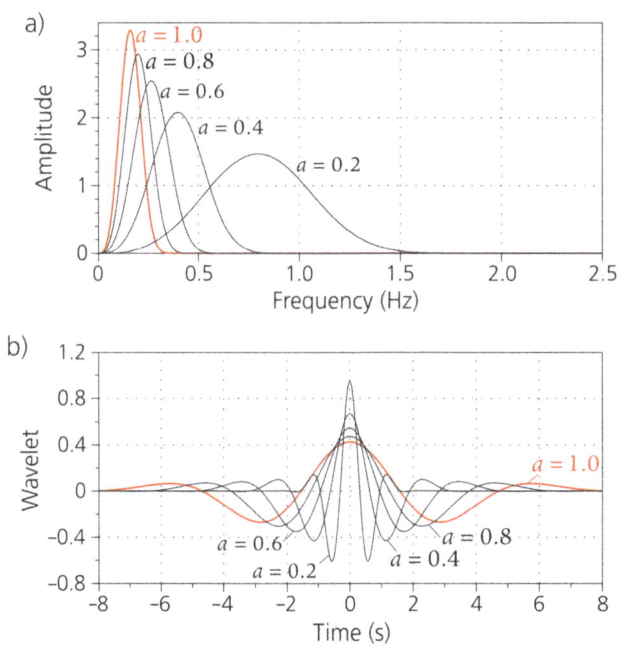

Figure 3.3 The standard zeroth-order Morse wavelet, defined with the parameters $(\beta, \gamma) = (3, 3)$. (a) The amplitude spectrum $\hat{\psi}^{(0)}_{\beta,\gamma}(f)$ of the zeroth-order Morse wavelet (red) and the spectra $\hat{\psi}_a(f) = \sqrt{a}\hat{\psi}^{(0)}_{\beta,\gamma}(af)$ of the dilated wavelets (black). (b) The real part of the Morse wavelet (red) and the dilated wavelets (black) in the time domain.

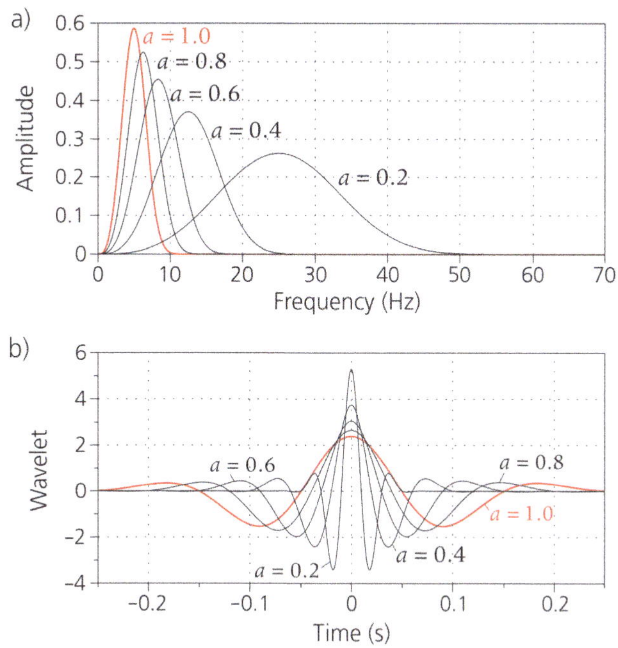

Figure 3.4 The modified zeroth-order Morse wavelet, defined with parameters $q = 5$ and $(\beta, \gamma) = (3, 3)$. (a) The amplitude spectrum $\hat{\psi}_q^{(0)}(f)$ of the modified Morse wavelet and the amplitude spectra $\sqrt{a}\,\hat{\psi}_q^{(0)}(af)$ of the dilated wavelets. (b) The real part of the modified Morse wavelet (red) and the dilated wavelets (black) in the time domain.

In the definition of the Morse wavelet in Eq. (3.19), the energy of the wavelet is normalised to unity, as required by the finite energy condition in Eq. (3.14). This definition differs from the conventional definition that normalises the maximum amplitude of the wavelet spectrum (Lilly & Olhede, 2009, 2010, 2012; Lilly, 2017).

Figure 3.3 displays the standard zeroth order Morse wavelet $\hat{\psi}_{\beta,\gamma}^{(0)}(f)$ (in red), defined with the parameters $(\beta, \gamma) = (3, 3)$. The dilated wavelets (in black) are defined with the dilation parameters $a = \{0.8, 0.6, 0.4, 0.2\}$, respectively. But even using such small dilation parameters, the dominant frequencies are still too low. The dominant frequency can be obtained, by setting $\mathrm{d}\hat{\psi}/\mathrm{d}f = 0$, as

$$f_c = \frac{1}{2\pi}\left(\frac{\beta}{\gamma}\right)^{1/\gamma}. \tag{3.20}$$

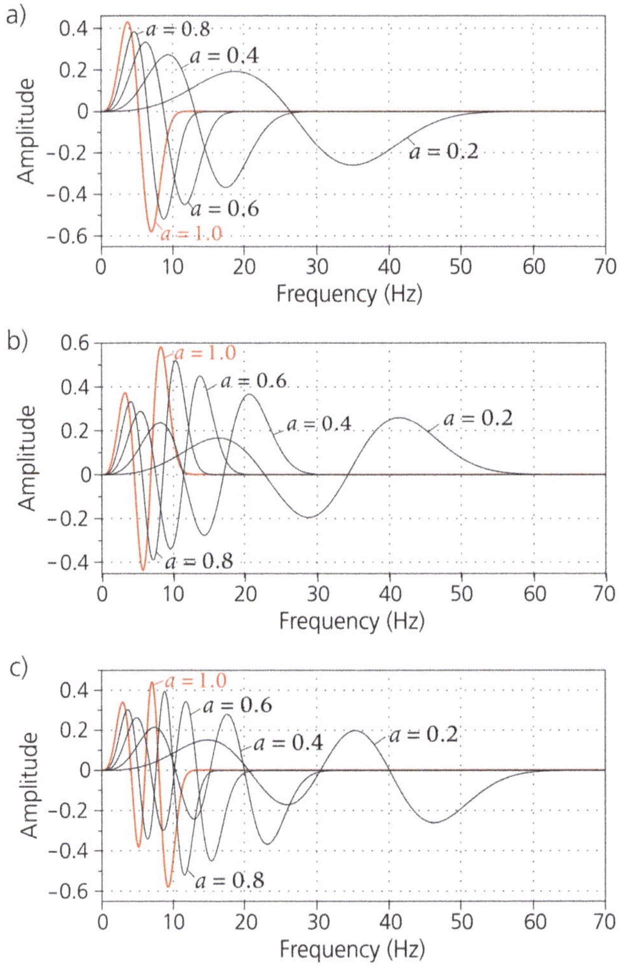

Figure 3.5 The frequency spectra of the modified Morse wavelets, defined with the parameters $q = 5$ and $(\beta, \gamma) = (3, 3)$. (a, b, c) The first, second and third order wavelets. The frequency spectra $\hat{\psi}(f)$ of the mother wavelets are shown in red, and the frequency spectra $\hat{\psi}_a(f)$ of the dilated wavelets in black.

The corresponding time-domain wavelets have a long duration. Therefore, the standard Morse wavelet is not suitable for the analysis of seismic signals. Here I propose to modify the zeroth order Morse wavelet as follows,

$$\hat{\psi}_q^{(0)}(f) = \sqrt{\frac{2^\alpha \gamma}{q\Gamma(\alpha)}} \left(\frac{f}{q}\right)^\beta e^{-\left(\frac{f}{q}\right)^\gamma}, \qquad \text{for } f \ge 0, \tag{3.21}$$

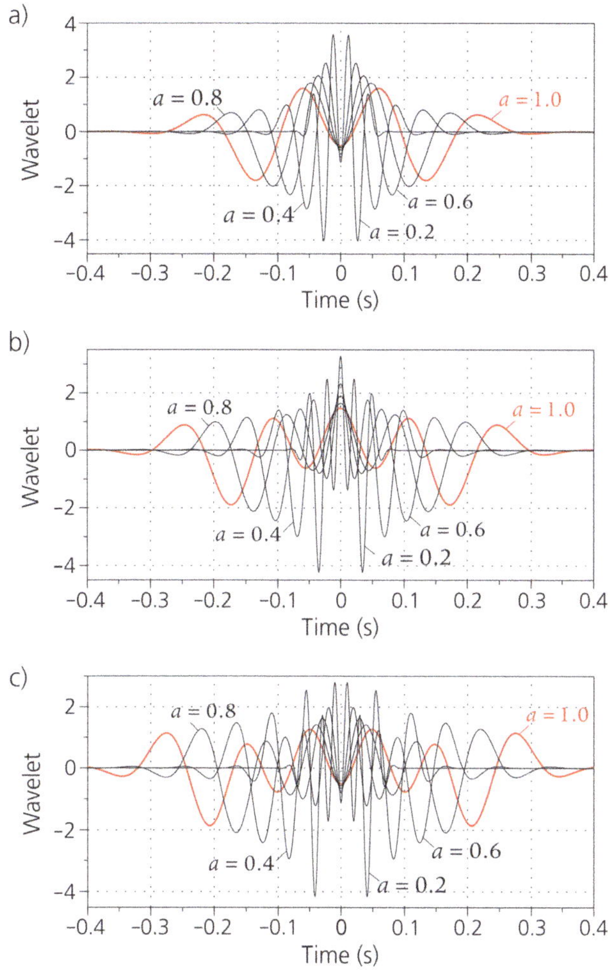

Figure 3.6 The modified Morse wavelets (real part) of the first (a), second (b), and third order (c). The mother wavelets $\psi(t)$ are shown in red, and the dilated wavelets $\psi_a(t)$ in black.

where q is a parameter adjusting the central frequency of the Morse wavelet:

$$f_c = q\left(\frac{\beta}{\gamma}\right)^{1/\gamma}. \tag{3.22}$$

In seismic signal analysis, the adjusting parameter q is determined from the peak frequency of the seismic spectrum, based on which an appropriate parameter pair (β, γ) is also assigned to the Morse wavelet.

Figure 3.4 shows the modified Morse wavelets with $q = 5$ and the same parameter pair (β, γ) as used in Figure 3.3. The frequency spectra are clearly different from those shown in Figure 3.3, especially the frequencies. Accordingly, the durations of the modified Morse wavelets also differ from those of the standard Morse wavelets.

In a similar way, I also modify the high-order Morse wavelets, as in the following expression:

$$\hat{\psi}_q^{(k)}(f) = \sqrt{\frac{k!2^\alpha \gamma}{q\Gamma(\alpha+k)}} \left(\frac{f}{q}\right)^\beta e^{-\left(\frac{f}{q}\right)^\gamma} \sum_{m=0}^{k} \frac{(-2)^m}{m!(k-m)!} \frac{\Gamma(\alpha+k)}{\Gamma(\alpha+m)} \left(\frac{f}{q}\right)^{m\gamma}, \quad \text{for } f \geq 0, \qquad (3.23)$$

where $k \geq 0$ is the order of the Morse wavelet.

Figure 3.5 shows the frequency spectra of the modified Morse wavelets with different orders $k = \{1, 2, 3\}$. The parameters defining these wavelets are $q = 5$ and $(\beta, \gamma) = (3, 3)$. The amplitudes of these frequency spectra have both positive and negative values. The total number of positive and negative peaks is related to the order of the wavelet by $k+1$.

The frequency spectra of the mother wavelets $\hat{\psi}(f)$ (with $a = 1.0$) are red, and the spectra of the dilated wavelets $\hat{\psi}_a(f)$ are black. The dilation parameters are $a = \{0.8, 0.6, 0.4, 0.2\}$, respectively. Figure 3.6 displays the modified Morse wavelets in the time domain. These Morse wavelets are complex wavelets that can be used in the continuous wavelet transform. Figure 3.6 displays only the real part of these wavelets.

3.4 The Generalised Seismic Wavelet

The Morlet wavelet has a symmetrical spectrum in the frequency domain and a symmetrical waveform in the time domain. The Morse wavelet can have an asymmetric spectrum in the frequency domain, depending on the parameter γ, but it also has a symmetric waveform in the time domain. The generalised seismic wavelet can better represent practically observed seismic signals that have asymmetric forms in both the time and frequency domains (Wang, 2015).

The Ricker wavelet is a well-known symmetric waveform in the time domain (Ricker, 1953). While the Ricker wavelet is the second derivative of a Gaussian function, generalisation of a seismic wavelet to an asymmetric form can be achieved by changing the order of the derivative from the integer '2' to a fractional value. For mathematical simplicity, the same Gaussian function used in the Ricker wavelet can be used as the basis function for the generalised seismic wavelet. Therefore, the non-Ricker, generalised seismic wavelets are systematically defined by fractional derivatives of the Gaussian function.

The generalised seismic wavelet in the frequency domain is defined as

$$\hat{\psi}^{(u)}(f) = \sqrt{\frac{2^{u+\frac{3}{2}}}{f_0 \Gamma\left(u+\frac{1}{2}\right)}} \left(\frac{f}{f_0}\right)^u \mathrm{e}^{-\frac{f^2}{f_0^2}} \mathrm{e}^{\mathrm{i}\pi\left(\frac{u}{2}+1\right)}, \qquad \text{for } f \geq 0, \tag{3.24}$$

where f_0 is a reference frequency (in Hz), and u is the parameter representing the spectrum apart from the symmetry $(u = 2)$. The reference frequency f_0 should physically reflect the anelastic property of the Earth's subsurface media, and thus is proportional to the standard deviation of the frequency spectrum of the generalised seismic wavelet.

According to Ricker (1943, 1944), a wavelet function (of the displacement, velocity, or acceleration type) can be expressed as a polynomial of various derivatives of a potential function. Therefore, we can put the potential function into a Gaussian form so that the derivative (of any order) is the Gaussian function multiplied by a polynomial. Here we represent the potential function as a negative Gaussian function:

$$g(t) = -\sqrt{\pi}\, f_0 \mathrm{e}^{-\pi^2 f_0^2 t^2}. \tag{3.25}$$

We represent the potential function as a negative Gaussian function and obtain the wavelets with a conventional polarity. The Fourier transform of the Gaussian function is

$$\hat{g}(f) = \int_{-\infty}^{\infty} g(t)\mathrm{e}^{-\mathrm{i}2\pi f t}\, \mathrm{d}t = -\mathrm{e}^{-\frac{f^2}{f_0^2}}. \tag{3.26}$$

The wavelet is defined as a fractional derivative of the Gaussian function with respect to time, $g^{(u)}(t) = \mathrm{d}^u g(t)/\mathrm{d}t^u$, where u is the order of the time-domain derivative. The fractional derivative is equivalent to a frequency factor $(\mathrm{i}2\pi f)^u$ in the frequency domain. Multiplying the frequency spectrum $\hat{g}(f)$ by the frequency factor,

$$(\mathrm{i}2\pi f)^u \hat{g}(f) = (2\pi f)^u \mathrm{e}^{-\frac{f^2}{f_0^2}} \mathrm{e}^{\mathrm{i}\pi\left(\frac{u}{2}+1\right)}, \tag{3.27}$$

and then normalising, we obtain the frequency-domain wavelet expression $\hat{\psi}^{(u)}(f)$ that is presented in Eq. (3.24).

If we perform the inverse Fourier transform of $\hat{\psi}^{(u)}(f)$, we get the wavelet $\psi^{(u)}(t)$ in the time domain. This generalised seismic wavelet is an analytic wavelet where the imaginary part is the Hilbert transform of the real part. Therefore, we only need to examine the real part of this analytic wavelet.

Figure 3.7 displays the amplitude spectra and the corresponding wavelets $\psi^{(u)}(t)$, defined with a reference frequency of $f_0 = 30$ Hz. The fractional parameter u varies between 0.5 and 3. For practical application, the parameter u can be statistically evaluated from the amplitude spectrum.

We are familiar with the derivatives of integer orders. For the fractional derivative $g^{(u)}(t)$, we can consider $\mathrm{d}^u/\mathrm{d}t^u$ as a transformation operator that continuously transforms $g(t)$ to $g^{(u)}(t)$ according to the continuously varying fractional parameter u (Appendix C).

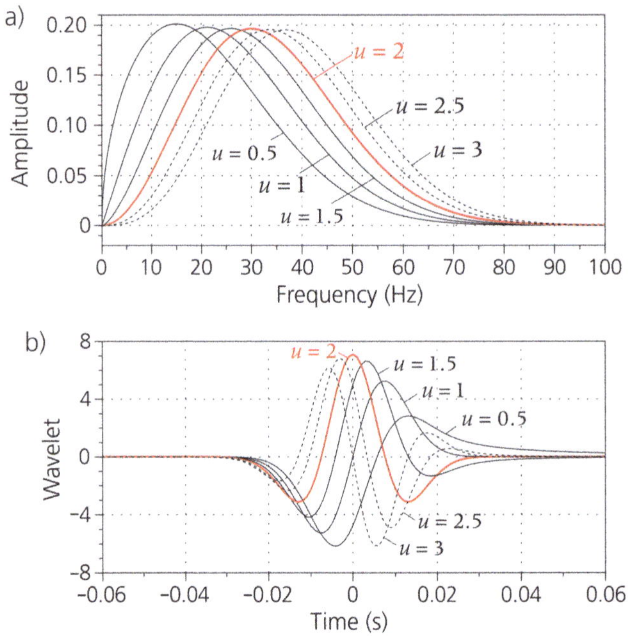

Figure 3.7 The generalised seismic wavelets, defined with the reference frequency $f_0 = 30$ Hz. (a) The amplitude spectra of the generalised seismic wavelets with different fractional factors between 0.5 and 3. (b) The real part of the generalised seismic wavelets in the time domain.

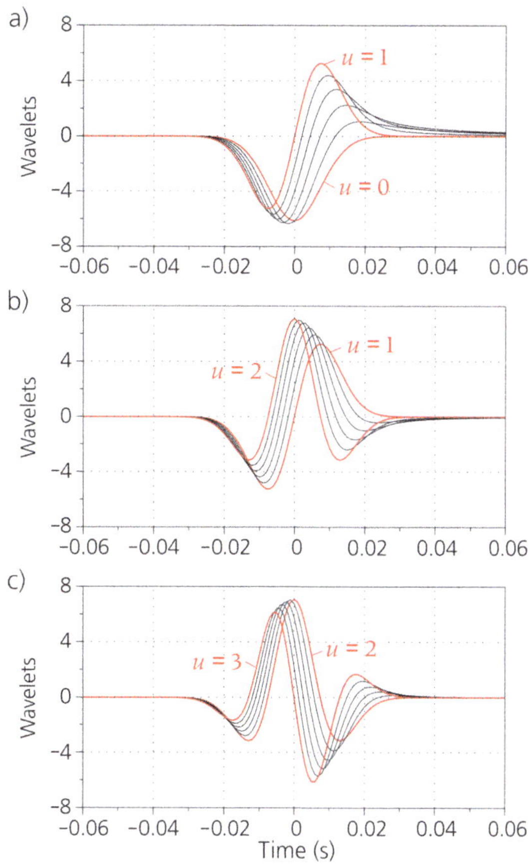

Figure 3.8 (a) The generalised seismic wavelets defined with fractional order $u = \{0, 0.2, 0.4, 0.6, 0.8, 1\}$. (b) The generalised seismic wavelets defined with fractional order $u = \{1, 1.2, 1.4, 1.6, 1.8, 2\}$. (c) The generalised seismic wavelets defined with fractional order $u = \{2, 2.2, 2.4, 2.6, 2.8, 3\}$. The red curves are those generated by integer-order derivatives. The reference frequency is $f_0 = 30$ Hz.

For instance, if $0 \le u \le 1$, the generalised seismic wavelets $g^{(u)}(t)$ vary continuously between $g^{(u=0)}(t)$ and $g^{(u=1)}(t)$, as displayed in Figure 3.8a, in which the zeroth order derivative is the function itself, $g^{(u=0)}(t) = g(t)$. The red curves are the wavelets generated by the integer-order derivatives.

In general, if $m-1 \le u \le m$, where m is an integer, the operator $\mathrm{d}^u / \mathrm{d}t^u$ transforms $g(t)$ to $g^{(u)}(t)$ between $g^{(m-1)}(t)$ and $g^{(m)}(t)$, and $g^{(m)}(t)$ can be given analytically. Figures 3.8b and 3.8c display the generalised seismic wavelets that are generated by fractional derivatives with fractional parameters varying between [1, 2] and [2, 3], respectively.

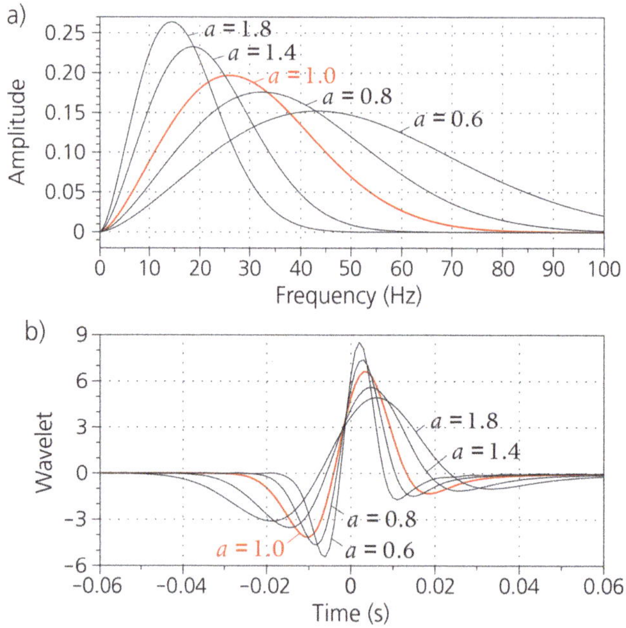

Figure 3.9 An example of the generalised seismic wavelet, defined with fractional order $u = 1.5$ and reference frequency $f_0 = 30$ Hz. (a) The amplitude spectra of the dilated wavelets with different dilation parameters. (b) The real part of the complex waveforms of the dilated wavelets.

Figure 3.9 shows an example of the generalised seismic wavelet $\psi^{(u)}(t)$, defined with $u = 1.5$ and $f_0 = 30\,\text{Hz}$, and the dilated versions. The dilation parameter a varies from 0.6 to 1.8. The time-domain wavelets are complex. The real parts of these dilated wavelets are close to the wavelets often observed in seismic field signals.

3.5 The Pseudo-frequency Representation

The continuous wavelet transform produces a two-dimensional spectrum in the time–scale plane, where the scale is the dilation parameter a (Rioul & Vetterli, 1991). However, in seismic signal analysis, a frequency representation is more intuitive than a scale, and the dilation parameter a is therefore converted to a frequency, to form a time–frequency spectrum. The converted frequency is a pseudo-frequency. The physical meaning of this pseudo-frequency is the mean frequency of a dilated wavelet.

The mean frequency f_m of a mother wavelet can be evaluated statistically as the first moment of power density in frequency,

$$f_m = \frac{\int_{-\infty}^{\infty} f\hat{\psi}^2(f)\mathrm{d}f}{\int_{-\infty}^{\infty} \hat{\psi}^2(f)\mathrm{d}f}. \tag{3.28}$$

For a dilated wavelet $\psi_a(t)$, the mean frequency f_a is

$$f_a = \frac{1}{a} f_m. \tag{3.29}$$

This relationship states that the pseudo-frequency of the dilated wavelet is inversely proportional to the dilation parameter $a > 0$ (Addison *et al.*, 2009). The relationship in Eq. (3.29) can be easily proved as

$$\begin{aligned} f_a &= \frac{\int_{-\infty}^{\infty} f\left[\sqrt{a}\,\hat{\psi}(af)\right]^2 \mathrm{d}f}{\int_{-\infty}^{\infty}\left[\sqrt{a}\,\hat{\psi}(af)\right]^2 \mathrm{d}f} \\ &= \frac{1}{a}\left(\frac{\int_{-\infty}^{\infty} \xi\hat{\psi}^2(\xi)\mathrm{d}\xi}{\int_{-\infty}^{\infty} \hat{\psi}^2(\xi)\mathrm{d}\xi}\right) = \frac{1}{a} f_m. \end{aligned} \tag{3.30}$$

For the frequency-domain Morlet wavelet $\hat{\psi}(f) = \hat{\psi}_\sigma(f)$ that is presented in Eq. 3.17, the characteristic peak frequency is f_c. The mean frequency of the mother wavelet is $f_m = f_c$. A dilated Morlet wavelet in the frequency domain is

$$\hat{\psi}_a(f) = \sqrt[4]{\frac{\pi\sigma^2 a^2}{(2\ln 2)f_c^2}}\,\mathrm{e}^{-\frac{\pi^2\sigma^2(af-f_c)^2}{(4\ln 2)f_c^2}}, \tag{3.31}$$

which has the mean frequency $f_a = f_m/a$.

The mean frequencies f_m and f_a are not necessarily the peak frequencies of the power spectra $\hat{\psi}^2(f)$ and $\hat{\psi}_a^2(f)$. For instance, the modified zeroth-order Morse wavelet $\hat{\psi}_q^{(0)}(f)$ has peak frequency f_c (Eq. 3.22), but its mean frequency is

$$f_m^{(0)} = \frac{q}{2^{1/\gamma}\Gamma(\alpha)}\Gamma\left(\alpha + \frac{1}{\gamma}\right). \tag{3.32}$$

For the modified first-order Morse wavelet, the frequency spectrum $\hat{\psi}_q^{(1)}(f)$ has two peaks (a positive peak and a negative peak, see Figure 3.5a):

$$f_c^{(1)} = q\left(\frac{1}{2}\left(1+\frac{1}{2\gamma}\right)+\frac{\beta}{\gamma}\pm\frac{1}{2}\sqrt{\left(1+\frac{1}{2\gamma}\right)^2+\frac{4\beta}{\gamma}}\right)^{1/\gamma}. \tag{3.33}$$

However, this mother wavelet has a unique mean frequency:

$$f_m^{(1)} = \frac{q}{2^{1/\gamma}\,\Gamma(\alpha+1)}\left(\alpha+\frac{1}{\gamma}+\frac{1}{\gamma^2}\right)\Gamma\left(\alpha+\frac{1}{\gamma}\right). \tag{3.34}$$

In general, for the modified kth-order Morse wavelet, the mean frequency $f_m^{(k)}$ is

$$f_m^{(k)} = \frac{k!q}{2^{1/\gamma}}\sum_{m=0}^{k}\sum_{n=0}^{k}\frac{(-1)^{m+n}}{m!n!(k-m)!(k-n)!}\frac{\Gamma(\alpha+k)}{\Gamma(\alpha+m)\Gamma(\alpha+n)}\Gamma\left(\alpha+\frac{1}{\gamma}+m+n\right). \tag{3.35}$$

For the generalised seismic wavelet, the analytical expression for the mean frequency is (Wang, 2015)

$$f_m = \frac{u\Gamma(u)}{\sqrt{2}\,\Gamma\left(u+\frac{1}{2}\right)}f_0. \tag{3.36}$$

In fact, not only the mean frequency, but also the frequency bandwidth, are inversely proportional to the dilation parameter a. The frequency bandwidth of a mother wavelet can be represented statistically by the standard deviation of the power spectrum:

$$f_{\sigma,m} = \left(\frac{\int_{-\infty}^{\infty}(f-f_m)^2\hat{\psi}^2(f)\mathrm{d}f}{\int_{-\infty}^{\infty}\hat{\psi}^2(f)\mathrm{d}f}\right)^{1/2}. \tag{3.37}$$

The frequency bandwidth is proportional to the standard deviation, and a rigorous relationship between these two quantities depends upon the type of wavelet.

For a dilated wavelet $\psi_a(t)$, the standard deviation $f_{\sigma,a}$ of the frequency spectrum $\hat{\psi}_a(f) = \sqrt{a}\,\hat{\psi}(af)$ is

$$f_{\sigma,a}=\frac{1}{a}f_{\sigma,m}. \tag{3.38}$$

This relationship can be easily proved as

$$\begin{aligned}
f_{\sigma,a} &= \left(\frac{\int_{-\infty}^{\infty}(f-f_a)^2\hat{\psi}_a^2(f)\mathrm{d}f}{\int_{-\infty}^{\infty}\hat{\psi}_a^2(f)\mathrm{d}f}\right)^{1/2} \\
&= \left(\frac{\int_{-\infty}^{\infty}(f-a^{-1}f_m)^2\left[\sqrt{a}\hat{\psi}(af)\right]^2\mathrm{d}f}{\int_{-\infty}^{\infty}\left[\sqrt{a}\hat{\psi}(af)\right]^2\mathrm{d}f}\right)^{1/2} \\
&= \frac{1}{a}\left(\frac{\int_{-\infty}^{\infty}(\xi-f_m)^2\hat{\psi}^2(\xi)\mathrm{d}\xi}{\int_{-\infty}^{\infty}\hat{\psi}^2(\xi)\,\mathrm{d}\xi}\right)^{1/2} = \frac{1}{a}f_{\sigma,m}.
\end{aligned} \tag{3.39}$$

A dilated wavelet with a larger dilation parameter $a>1$ has a lower mean frequency and a narrower frequency band, so that the time–frequency analysis achieves a higher resolution on the frequency axis. Conversely, a dilated wavelet with a smaller dilation parameter $0<a<1$ has a higher mean frequency and a wider frequency band, corresponding to a smaller width in the time domain, so that the time–frequency analysis shows more detail in the time axis.

3.6 The Inverse Wavelet Transform

Because the continuous wavelet transform is a redundant transform, there is no unique way to define the inverse wavelet transform. The classical inversion formula for the continuous wavelet transform is expressed as deconvolution in the time domain (Holschneider, 1995; Liu & Hsu, 2012):

$$x(t)=\frac{1}{C_\psi}\int_{-\infty}^{\infty}\int_{-\infty}^{\infty}S_{\mathrm{CWT}}(\tau,f)I(t-\tau,f)\mathrm{d}\tau\mathrm{d}f, \tag{3.40}$$

where $I(t,f)$ is the inversion kernel, and C_ψ is called an admissible constant that satisfies the admissibility condition:

$$C_\psi=\int_{-\infty}^{\infty}\hat{\psi}^*(\xi,f)\hat{I}(\xi,f)\mathrm{d}f=\text{constant}\neq 0. \tag{3.41}$$

Here, $\hat{\psi}(\xi, f)$ is the Fourier transform of $\psi(t, f)$ with respect to time t, and $\hat{I}(\xi, f)$ is the Fourier transform of $I(t, f)$ with respect to time t.

To prove the classic inversion formula in Eq. (3.40), let us denote $u = t - \tau,\ v = t - t'$, and

$$\rho(v) = \int_{-\infty}^{\infty}\int_{-\infty}^{\infty} \psi^*(u - v, f) I(u, f) \mathrm{d}u \mathrm{d}f. \tag{3.42}$$

Note that

$$\hat{\rho}(\xi) = \int_{-\infty}^{\infty} \hat{\psi}^*(\xi, f) \hat{I}(\xi, f) \mathrm{d}f. \tag{3.43}$$

If Eq. (3.41) holds, one can have that

$$\rho(v) = C_\psi \delta(v); \tag{3.44}$$

thus, the right-hand side of Eq. (3.40) is

$$\begin{aligned}
\tilde{x}(t) &= \frac{1}{C_\psi} \int_{-\infty}^{\infty}\int_{-\infty}^{\infty} \left(\int_{-\infty}^{\infty} x(t')\psi^*(t' - \tau, f)\mathrm{d}t' \right) I(t - \tau, f)\mathrm{d}\tau \mathrm{d}f \\
&= \frac{1}{C_\psi} \int_{-\infty}^{\infty} x(t')\mathrm{d}t' \int_{-\infty}^{\infty}\int_{-\infty}^{\infty} \psi^*(t' - \tau, f) I(t - \tau, f)\mathrm{d}\tau \mathrm{d}f \\
&= \frac{1}{C_\psi} \int_{-\infty}^{\infty} x(t - v)\mathrm{d}v \int_{-\infty}^{\infty}\int_{-\infty}^{\infty} \psi^*(u - v, f) I(u, f)\mathrm{d}u \mathrm{d}f \\
&= \frac{1}{C_\psi} \int_{-\infty}^{\infty} x(t - v)\rho(v)\mathrm{d}v = \int_{-\infty}^{\infty} x(t - v)\delta(v)\mathrm{d}v = x(t).
\end{aligned} \tag{3.45}$$

A practical approach to the inverse wavelet transform can be formed by considering $f = f_m / a$ and setting

$$I(t, f) = \psi(t, f) = \psi\left(t, \frac{1}{a}\right) = \frac{1}{\sqrt{|a|}} \psi\left(\frac{t}{a}\right), \tag{3.46}$$

which is equivalently in the frequency domain,

$$\hat{I}(\xi, f) = \hat{\psi}(\xi, f) = \hat{\psi}\left(\xi, \frac{1}{a}\right) = \sqrt{|a|}\, \hat{\psi}(a\xi). \tag{3.47}$$

The admissibility condition of Eq. (3.41) becomes

$$\begin{aligned} C_\psi &= -f_m \int_{-\infty}^{\infty} \left| \hat{\psi}\left(\xi, \frac{1}{a}\right) \right|^2 \frac{\mathrm{d}a}{a^2} \\ &= -f_m \int_{-\infty}^{\infty} |\hat{\psi}(a\xi)|^2 \frac{\mathrm{d}a}{|a|} \\ &= -f_m \int_{-\infty}^{\infty} |\hat{\psi}(u)|^2 \frac{1}{|u|} \mathrm{d}u = \text{constant} \neq 0. \end{aligned} \tag{3.48}$$

Therefore, the admissibility condition is

$$C_\psi = \int_{-\infty}^{\infty} \frac{|\hat{\psi}(f)|^2}{|f|} \mathrm{d}f < \infty. \tag{3.49}$$

This admissibility condition implies that $\hat{\psi}(f = 0) = 0$, which is equivalent to the primary condition of Eq. (3.13): the time integral of the wavelet $\psi(t)$ is zero. Such a conventional wavelet is called an admissible wavelet. With such an admissible wavelet $\psi(t)$, the inverse wavelet transform can be written, using the inversion kernel of Eq. (3.47), as follows:

$$x(t) = \frac{1}{C_\psi} \int_{-\infty}^{\infty} \mathrm{d}\tau \int_{0}^{\infty} \frac{1}{a^2} S_{\text{CWT}}(\tau, a) \psi_a(t - \tau) \mathrm{d}a. \tag{3.50}$$

3.7 Implementation of the Continuous Wavelet Transform

The continuous wavelet transform can be interpreted as a time convolution between the seismic signal and the time-reversed dilated wavelets, which provide temporal specificity. Therefore, the continuous wavelet transform in the frequency domain can be interpreted as a bandpass filter bank (Rioul & Duhamel, 1992; Vetterli & Herley, 1992) and can be implemented by Fourier transforms:

(1) Performing the Fourier transform of seismic signal $x(t)$, and denoting it as $\hat{x}(\xi)$.
(2) Setting the parameters for the mother wavelet, building the mother wavelet in the frequency domain, and determining the mean frequency f_m.

(3) Deriving the dilation parameter $a = f_m/f$, corresponding to any given pseudo-frequency f.
(4) Calculating the Fourier transform $\hat{\psi}_a(\xi) = \sqrt{a}\,\hat{\psi}(a\xi)$ of the dilated wavelet with $a > 0$.
(5) Generating the time–frequency spectrum by the inverse Fourier transform,

$$S_{\text{CWT}}(t, f) = \mathcal{F}_{\xi}^{-1}\left[\hat{x}(\xi)\hat{\psi}_a^*(\xi)\right]. \tag{3.51}$$

Steps 3–5 are repeated. At each iteration, these three steps calculate the 'time' spectrum for a given pseudo-frequency $f = f_a > 0$.

The Fourier transforms in the above implementation are complex-to-complex transforms. Even if the original signal $x(t)$ is real, the Fourier transform treats it as a complex signal, where the imaginary part is zero. The periodicity of the signal outside the observation interval is assumed by virtue of the fast Fourier transform (FFT). The final time–frequency spectrum is complex and contains the phase information.

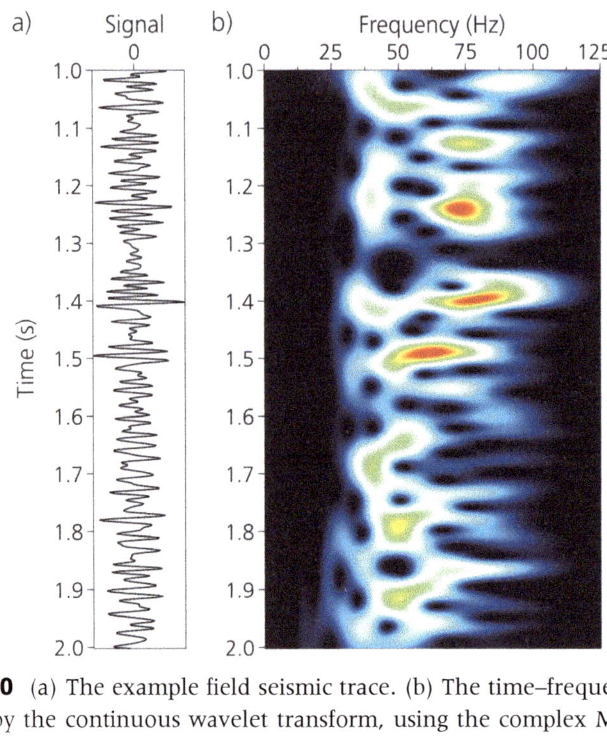

Figure 3.10 (a) The example field seismic trace. (b) The time–frequency spectrum generated by the continuous wavelet transform, using the complex Morlet wavelet defined with parameters $\sigma = 2.5$ and $f_c = 30$ Hz.

Figure 3.10 is an example of the continuous wavelet transform, implemented through forward and inverse Fourier transforms. The mother wavelet is the complex Morlet wavelet with the spread parameter $\sigma = 2.5$ and the characteristic frequency $f_c = 30$ Hz. The Fourier transform of the Morlet wavelet $\hat{\psi}(\xi)$ is given by Eq. (3.17), and the dilated wavelet $\hat{\psi}_a(\xi)$ with $a > 0$ is explicitly given by Eq. (3.31).

The inverse wavelet transform (Eq. 3.50) can also be implemented efficiently through the forward and inverse Fourier transforms:

$$x(t) = \frac{1}{C_\psi} \int_0^\infty \frac{1}{a^2} \mathcal{F}_\xi^{-1} \left[\hat{S}_{\mathrm{CWT}}(\xi, a) \hat{\psi}_a(\xi) \right] \mathrm{d}a, \tag{3.52}$$

where $\hat{S}_{\mathrm{CWT}}(\xi, a) = \mathcal{F}_t[S_{\mathrm{CWT}}(t, a)]$, for $a > 0$, is the Fourier transform of the time–frequency spectrum $S_{\mathrm{CWT}}(t, a)$ with respect to time t. There is a singularity issue theoretically, when $a \to 0$ in the continuous wavelet transform. In practice, the dilation parameter a has a lower bound as $a \ge f_m / f_{\mathrm{Nyq}}$, if the spectrum is restrained by the Nyquist frequency f_{Nyq}.

3.8 Hydrocarbon Reservoir Characterisation

In the previous chapter, we discussed the inverse Q filtering on the time–frequency spectrum produced by the Gabor transform. We can also easily apply the same inverse Q filter, straightforwardly, on the time–frequency spectrum produced by the continuous wavelet transform (Braga & Moraes, 2013). In this section, we demonstrate the application of the time–frequency spectrum produced by the continuous wavelet transform in the characterisation of hydrocarbon reservoirs.

The geophysical objective of the application example is to detect the sand distribution within a target layer. Figure 3.11a shows the depth structure of the target horizon. The overall structure appears to be simple and complete. There are two high and steep anticlines, with a broad, gentle syncline in between, showing a spread in a north-northeast direction. The local structure is not pronounced, and the depth contour is only affected by some minor faults.

Between the two hills there is a long and narrow valley, which spreads in a north-northeast direction. We try to use the time–frequency spectrum produced by the continuous wavelet transform to identify the sand distribution in the valley.

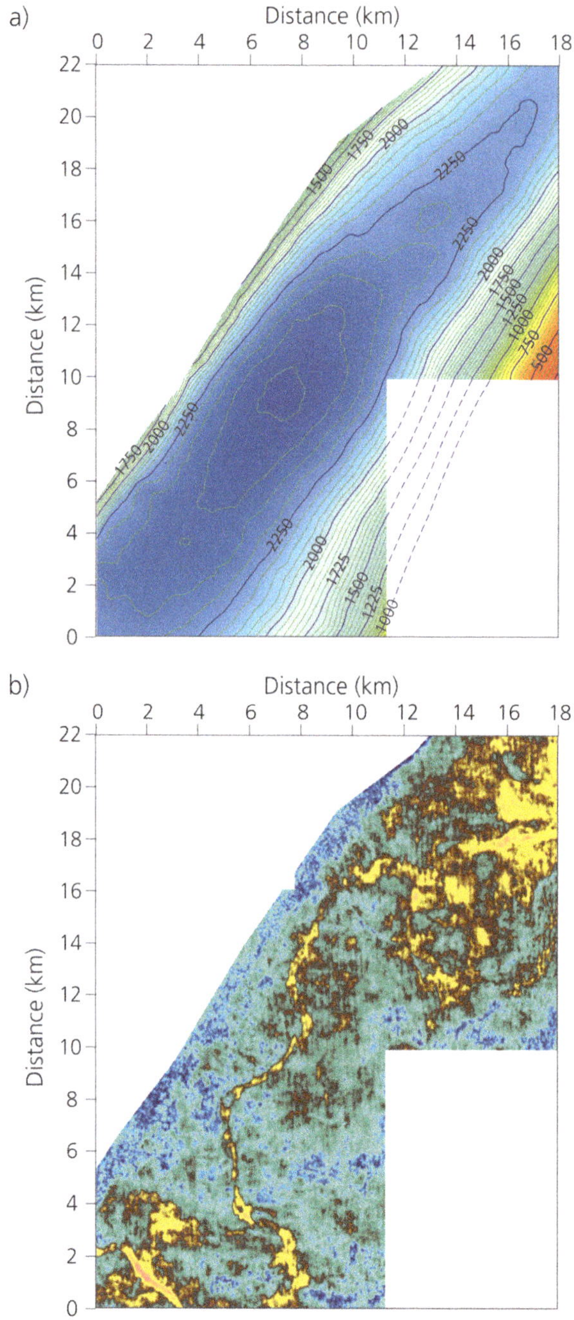

Figure 3.11 The depth structure (a) and an iso-frequency spectral slice (b) of a target horizon. The contours in the structure indicate the depth (in m), and the spectral slice at 52 Hz shows the sand distribution within a deep valley.

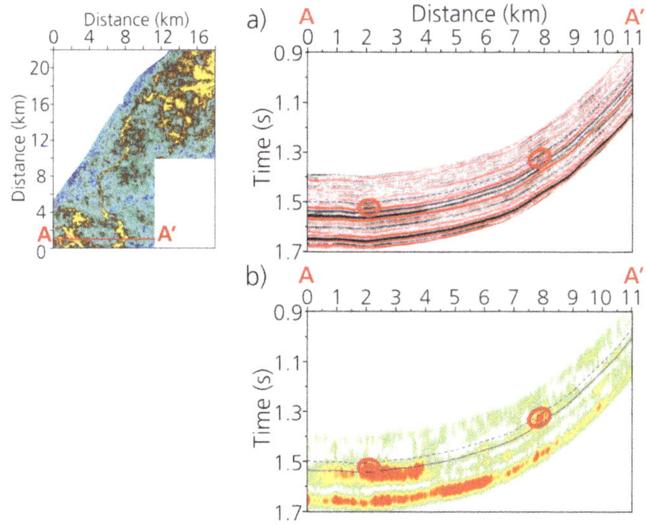

Figure 3.12 The sand channels (ellipses) within the target layer. (a) An example seismic profile A–A′, showing two waveform anomalies (marked by ellipses) corresponding to two sand channels. (b) The corresponding spectral profile of an iso-frequency at 52 Hz.

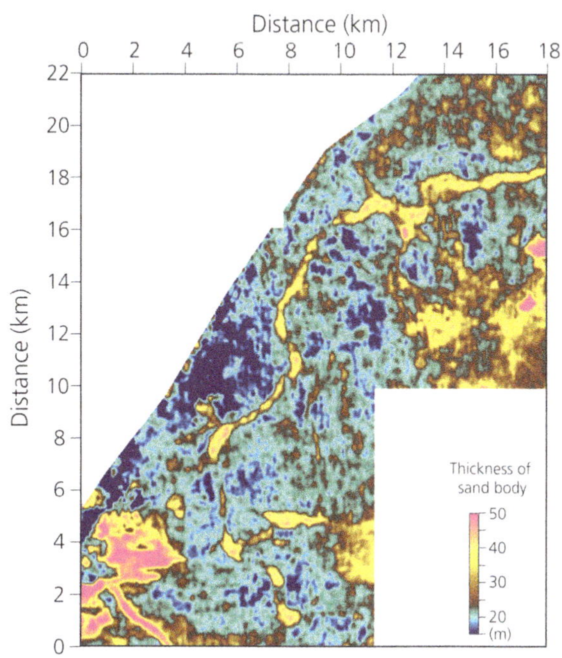

Figure 3.13 The estimated thickness of the sand bodies distributed over the target layer.

The continuous wavelet transform is defined as the cross-correlation between the seismic trace and the dilated versions, at different timescales, of a basic wavelet function with a zero mean. Therefore, each sample in the time–frequency spectrum is a coefficient, indicating how well the local seismic segment, at each instant of time, correlates with each dilated wavelet. The time–frequency analysis of the seismic data shows that the dominant frequency of the sand reflections is about 52 Hz.

Figure 3.11b shows the amplitude spectral slice along the horizon, extracted from the 3D spectral cube of the iso-frequency at 52 Hz. This spectral slice clearly shows the sand distribution within the deep valley. There is a ribbon-like sand channel that extends across the entire horizon. In addition, there are fluvial fans in the south and extensive alluvial fans in the north, mixed with braided streams.

Figure 3.12a shows a seismic profile (the line A–A′ marked in the spectral slice). On the seismic reflection profile, the target layer interval is marked with a solid curve and a dashed curve. The seismic waveforms of the two sand channels appear as anomalies in the profile.

Figure 3.12b shows the corresponding iso-frequency profile at a frequency of 52 Hz. Two sand channels are clearly visible in the iso-frequency profile and marked by the ellipses. The spectral profile also shows that the left sand channel coexists with fluvial fans.

Finally, Figure 3.13 shows the estimated thickness of the sand bodies distributed over the target layer. The thickness (in m) is estimated based on the spectral amplitudes of several iso-frequencies and the impedances obtained from the seismic inversion.

4

The S Transform

In the previous chapters we discussed two methods, the Gabor transform and the continuous wavelet transform, for time–frequency analysis of a nonstationary seismic trace. The Gabor transform uses a sliding window to segment a seismic trace prior to initiating the Fourier transform. Thus, the frequency spectrum at a time position reflects the local characteristics of the seismic signal. However, since the window function used in the Gabor transform has a fixed window length, the number of cycles of oscillation within the window varies along the frequency axis. Consequently, the resolution of the local spectrum at high frequencies is relatively lower than the resolution at low frequencies, at the same time position (Ventosa *et al.*, 2008).

The continuous wavelet transform introduces a timescale variable, the dilation parameter, to dilate and contract a basis function (the mother wavelet) to have a fixed number of cycles within the envelope of the scaled functions (the daughter wavelets). The pseudo-frequency in the wavelet-transform spectrum is inversely proportional to the dilation parameter used to dilate and contract the basis function. Therefore, the resolution of the wavelet-transform spectrum remains constant along the pseudo-frequency axis.

This chapter deals with the S transform, which can be seen as an intermediate method between the Gabor transform and the continuous wavelet transform. The S transform defines the window as a function of frequency. Such a frequency-dependent window definition enables the S transform to have a fixed number of cycles within a window and to achieve a constant resolution along the frequency axis.

Time–Frequency Analysis of Seismic Signals, First Edition. Yanghua Wang.

The definition of the S transform is similar to the definition of the Gabor transform, as both transform methods define the window function in a Gaussian form. The S transform can also be considered a special case of the continuous wavelet transform. The 'wavelet' in this special case is defined as a complex harmonic function enveloped by a Gaussian window, where the size of the Gaussian window varies continuously with frequency.

4.1 Basics of the S Transform

The S transform (ST) of time signal $x(t)$ is expressed as

$$S_{\mathrm{ST}}(t,f) = \int_{-\infty}^{\infty} x(\tau)w(t-\tau,f)\mathrm{e}^{-\mathrm{i}2\pi f\tau}\mathrm{d}\tau, \tag{4.1}$$

where $w(t,f)$ is the Gaussian window function that is defined in the time–frequency plane, and t indicates the central position of the sliding window along the time axis. A distinguishing feature of the S transform is that the Gaussian window function is frequency dependent.

The S transform can be viewed as an extension of the Gabor transform. In the Gabor transform, the sliding window function is defined in the Gaussian form (Eq. 2.12):

$$w(t) = \frac{1}{\sqrt[4]{\pi\sigma^2}}\mathrm{e}^{-\frac{t^2}{2\sigma^2}}. \tag{4.2}$$

The Gaussian spread parameter σ controls the window breadth and thus controls the number of signal oscillations falling within the window. In the definition of the S transform (Stockwell *et al.*, 1996), the Gaussian spread parameter σ is a function of frequency:

$$\sigma = \frac{k}{|f|}, \qquad f \neq 0, \tag{4.3}$$

where $k > 0$ is a constant. The frequency-dependency of the Gaussian spread parameter σ maintains a constant number of signal oscillations falling within the window. The constant k permits specific time and frequency resolution on the time–frequency plane (Mansinha *et al.*, 1997; Pinnegar & Mansinha, 2003; Simon *et al.*, 2007).

Therefore, the Gaussian window function that is used in the S transform is a two-dimensional (2D) function,

$$w(t,f)=\sqrt{\frac{|f|}{k\sqrt{\pi}}}\mathrm{e}^{-\frac{f^2t^2}{2k^2}}, \tag{4.4}$$

and the S transform is written in the form of a time integral, as

$$S_{\mathrm{ST}}(t,f)=\sqrt{\frac{|f|}{k\sqrt{\pi}}}\int_{-\infty}^{\infty}x(\tau)\mathrm{e}^{-\frac{f^2(t-\tau)^2}{2k^2}}\mathrm{e}^{-\mathrm{i}2\pi f\tau}\mathrm{d}\tau. \tag{4.5}$$

The 2D Gaussian window function $w(t,f)$ is normalised, so that the energy becomes unity, $\int_{-\infty}^{\infty}|w(t,f)|^2\mathrm{d}t=1$. The normalisation is consistent with the treatment in the previous chapters. According to Parseval's theorem (Eq. 1.21), the normalised window function does not affect the spectral energy of the signal. For a real-valued seismic signal $x(t)$, only non-negative frequencies are considered, and the energy of the time–frequency spectrum is

$$\int_{-\infty}^{\infty}|x(t)|^2\mathrm{d}t=2\int_{0}^{\infty}\mathrm{d}f\int_{-\infty}^{\infty}|S_{\mathrm{ST}}(t,f)|^2\,\mathrm{d}t. \tag{4.6}$$

However, once the S transform satisfies Parseval's theorem, as shown in Eq. (4.6), the time integral of the S transform $S_{\mathrm{ST}}(t,f)$ does *not* represent the local frequency spectrum, because the time integral is not equal to the Fourier spectrum of the signal $x(t)$:

$$\begin{aligned}\int_{-\infty}^{\infty}S_{\mathrm{ST}}(t,f)\mathrm{d}t&=\int_{-\infty}^{\infty}\sqrt{\frac{|f|}{k\sqrt{\pi}}}\int_{-\infty}^{\infty}x(\tau)\mathrm{e}^{-\frac{|f|^2(t-\tau)^2}{2k^2}}\mathrm{e}^{-\mathrm{i}2\pi f\tau}\mathrm{d}\tau\mathrm{d}t\\&=\int_{-\infty}^{\infty}\sqrt{\frac{|f|}{k\sqrt{\pi}}}\mathrm{e}^{-\frac{|f|^2(t-\tau)^2}{2k^2}}\mathrm{d}t\int_{-\infty}^{\infty}x(\tau)\mathrm{e}^{-\mathrm{i}2\pi f\tau}\mathrm{d}\tau\\&=\sqrt{\frac{2k\sqrt{\pi}}{|f|}}\hat{x}(f),\end{aligned} \tag{4.7}$$

where $\hat{x}(f)$ is the Fourier transform of $x(t)$. This feature, $\int_{-\infty}^{\infty}S_{\mathrm{ST}}(t,f)\mathrm{d}t\neq\hat{x}(f)$, significantly differs from the conventional definition of the S transform

(Stockwell *et al.*, 1996; Stockwell, 2007), in which $\int_{-\infty}^{\infty} S_{\mathrm{ST}}(t,f)\mathrm{d}t = \hat{x}(f)$ is assumed, but $\int_{-\infty}^{\infty} |w(t,f)|^2 \mathrm{d}t \neq 1$, the energy of the frequency-dependent Gaussian window is not unity.

The physical meaning of the S transform can be understood as a time convolution process, at any given frequency f, between the complex signal $x(t)\mathrm{e}^{-\mathrm{i}2\pi ft}$ and the Gaussian window function:

$$S_{\mathrm{ST}}(t,f) = x(t)\mathrm{e}^{-\mathrm{i}2\pi ft} * w(t,f), \tag{4.8}$$

where $*$ is the convolution operator with respect to time t. The Fourier transform of the complex signal $x(t)\mathrm{e}^{-\mathrm{i}2\pi ft}$ can be represented as

$$\hat{x}(\xi + f) = \mathcal{F}_t\left[x(t)\mathrm{e}^{-\mathrm{i}2\pi ft}\right], \tag{4.9}$$

where $\mathcal{F}_t$ is the Fourier transform with respect to time t, and ξ is the corresponding frequency.

The Gaussian window function $w(t,f)$ is symmetrical in time and frequency, and the Fourier transform $\hat{w}(\xi,f)$ is also in a Gaussian shape. Moreover, this Fourier transform $\hat{w}(\xi,f)$ has no sidelobes in the 2D space, as shown in Figure 2.9c.

In the Fourier transform domain, the S transform can be written explicitly as

$$\begin{aligned} S_{\mathrm{ST}}(t,f) &= \mathcal{F}_\xi^{-1}\left[\hat{x}(\xi+f)\hat{w}(\xi,f)\right] \\ &= \sqrt{\frac{2k\sqrt{\pi}}{|f|}} \int_{-\infty}^{\infty} \hat{x}(\xi+f)\mathrm{e}^{-2\left(\frac{\pi k\xi}{f}\right)^2} \mathrm{e}^{\mathrm{i}2\pi\xi t}\mathrm{d}\xi, \quad f \neq 0. \end{aligned} \tag{4.10}$$

The singularity in Eq. (4.10) indicates, intuitively, that there is a sufficient condition for the S transform:

$$\int_{-\infty}^{\infty} x(t)\mathrm{d}t = 0. \tag{4.11}$$

That is, the seismic signal to be analysed has a zero mean. Consequently, the S transform sets $S_{\mathrm{ST}}(t,f) = 0$, for $f = 0$.

In the numerical implementation with a finite time series, if using Eq. (4.10) with the Fourier transform $\hat{x}(\xi+f)$, note that the frequency spectrum $\hat{x}(\xi+f)$ includes frequency components that are outside the Nyquist

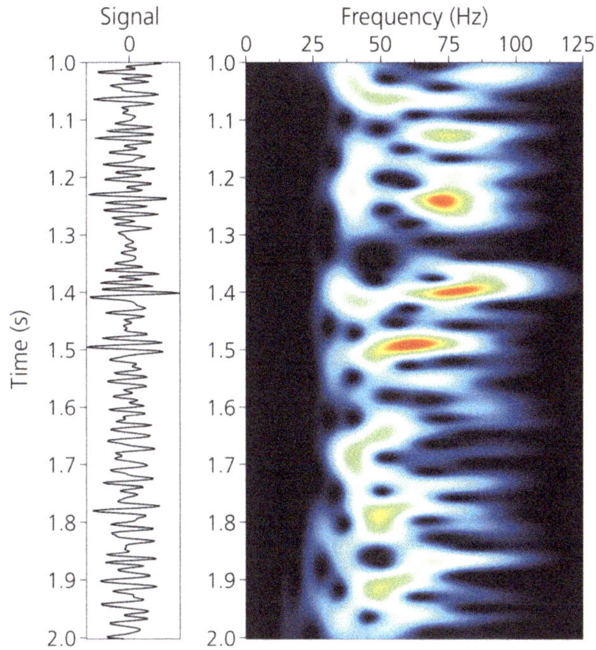

Figure 4.1 The time–frequency spectrum produced by the standard S transform, using the Fourier transform-based method.

frequency. Thus, Eq. (4.10) suffers from an aliasing issue, because of the periodicity assumption that is implicitly made in any discrete Fourier transform.

Figure 4.1 displays an S transform example, in which $k = 1$ is used, for the sake of comparison to the Gabor transform that were shown in Figures 2.13 and 2.14. The time–frequency spectrum indicates that the S transform, which uses a frequency-dependent window function, provides good localisation in the time axis for higher frequencies. However, it provides low resolution in the time axis for low frequencies.

4.2 The Generalised S Transform

To achieve an optimal spectral representation in the time–frequency plane, various modifications to the standard S transform method have been developed. As stated in Eq. (4.8), the standard S transform, at any given frequency f, is a time convolution of the complex signal $y(t, f) = x(t)e^{-i2\pi ft}$

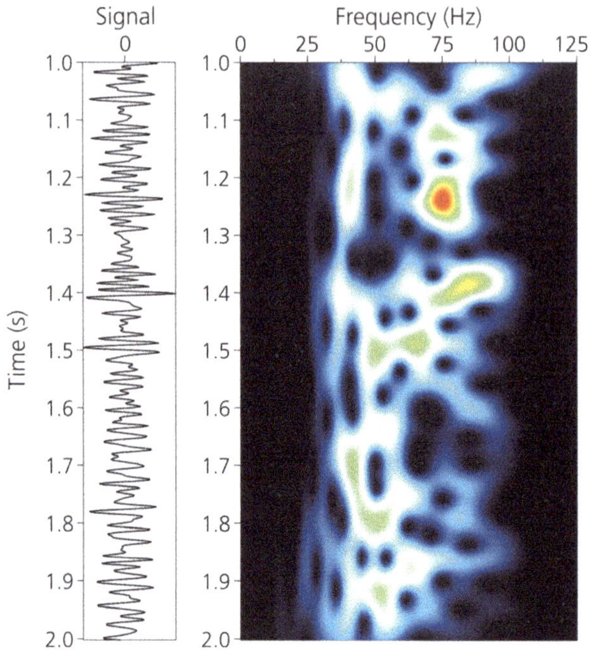

Figure 4.2 The time–frequency spectrum produced by the generalised S transform, with parameters ($k = 1$, $p = 0.85$). The method effectively reduces spectral smearing on the frequency axis at high frequencies, but at the same time increases spectral smearing on the time axis at low frequencies.

with the Gaussian window function $w(t, f)$. Therefore, modifying either of these two parts can lead to variants of the S transform.

The generalised S transform is one variant of the S transform, which modifies the Gaussian window function $w(t, f)$, by replacing the frequency $|f|$ in the spread parameter σ with a power of frequency:

$$\sigma = \frac{k}{|f|^p}, \qquad f \neq 0, \tag{4.12}$$

where $p > 0$ is a power parameter for tuning the frequency-dependency of the window width (Djurović *et al.*, 2008; Sejdić *et al.*, 2008, 2009). The 2D Gaussian window function becomes

$$w_p(t, f) = \sqrt{\frac{|f|^p}{k\sqrt{\pi}}} e^{-\frac{|f|^{2p} t^2}{2k^2}}. \tag{4.13}$$

Thus, the generalised S transform (GST) can be written in the form of a time integral as

$$S_{\text{GST}}(t,f;p)=\sqrt{\frac{|f|^{p}}{k\sqrt{\pi}}}\int_{-\infty}^{\infty}x(\tau)\mathrm{e}^{-\frac{|f|^{2p}(t-\tau)^{2}}{2k^{2}}}\mathrm{e}^{-\mathrm{i}2\pi f\tau}\mathrm{d}\tau. \tag{4.14}$$

In the generalised S transform, the power parameter p is an extra parameter for tuning the window breadth according to the frequency. Consequently, the time-averaged frequency spectrum also depends on the frequency raised to the power of p:

$$\int_{-\infty}^{\infty}S_{\text{GST}}(t,f;p)\mathrm{d}t=\sqrt{\frac{2k\sqrt{\pi}}{|f|^{p}}}\hat{x}(f)\neq\hat{x}(f). \tag{4.15}$$

Figure 4.2 displays an example time–frequency spectrum generated by the generalised S transform, which is defined by parameters $k=1$ and $p=0.85$. A smaller p value $(0<p<1)$ in the generalised S transform effectively reduces the frequency-axis spectral smearing at high frequencies but simultaneously increases the time-axis spectral smearing at low frequencies.

4.3 The Fractional Fourier Transform

The Fourier transform kernel $\mathrm{e}^{-\mathrm{i}2\pi ft}$ represents a phase rotation. In the standard and generalised S transform, the Fourier transform kernel $\mathrm{e}^{-\mathrm{i}2\pi ft}$ is applied to the original signal $x(t)$, to form a complex signal $y(t,f)=x(t)\mathrm{e}^{-\mathrm{i}2\pi ft}$, before the complex signal is convolved with a window function. Therefore, replacing the ordinary Fourier transform kernel with a fractional Fourier transform kernel, in forming the complex signal, will lead to another variant of the S transform. Before looking at its potential application of the fractional Fourier transform kernel, this section summarises the basics of the fractional Fourier transform.

For the fractional Fourier transform, the forward and inverse transform pair can be expressed as

$$\hat{x}_{\alpha}(u)=\mathcal{F}_{\alpha}\left[x(t)\right]=\int_{-\infty}^{\infty}x(t)K_{\alpha}(t,u)\mathrm{d}t, \tag{4.16}$$

$$x(t) = \mathcal{F}_\alpha^{-1}\left[\hat{x}_\alpha(u)\right] = \int_{-\infty}^{\infty} \hat{x}_\alpha(u) K_\alpha^*(t,u) \mathrm{d}u, \tag{4.17}$$

where α is a fractional value, u is an auxiliary frequency parameter, $K_\alpha(t,u)$ is the fractional Fourier transform kernel, and $K_\alpha^*(t,u)$, the complex conjugate of $K_\alpha(t,u)$, is the inverse fractional Fourier transform kernel.

The physical meaning of the fractional value α is an angle of phase rotation in the time–frequency plane (Namias, 1980; McBride & Kerr, 1987; Almeida, 1994). The ordinary Fourier transform can be viewed as a rotation from the time axis to the frequency axis. Since the time and frequency axes are orthogonal, the rotation angle is $\alpha = \pi/2$.

Figure 4.3 illustrates that the fractional Fourier transform, $x(t) \rightarrow \hat{x}_\alpha(u)$, is a rotation from the time axis t to an axis u with an arbitrary angle α. In the context of the S transform, the variation range of the rotation angle is set as $0 < \alpha \leq \pi/2$.

The axis u in the fractional Fourier transform is an auxiliary frequency parameter. It is not the ordinary frequency f but rather it is a projection $u = f \sin\alpha$. When $\alpha = \pi/2$, then $u = f$, which is the ordinary frequency.

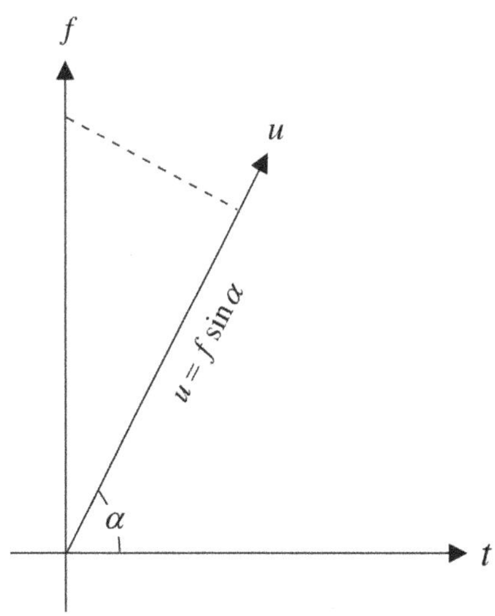

Figure 4.3 The fractional Fourier transform $x(t) \rightarrow \hat{x}_\alpha(u)$ is a rotation from the time axis t to an auxiliary axis u with an arbitrary angle α. The ordinary Fourier transform can be considered as a rotation from the time axis t to the frequency axis f, and the angle of rotation is therefore $\alpha = \pi/2$. The auxiliary axis u is a projected frequency, $u = f \sin\alpha$.

The fractional Fourier transform kernel $K_\alpha(t,u)$ is defined as

$$K_\alpha(t,u) = \sqrt{\csc\alpha}\,\mathrm{e}^{-\mathrm{i}\frac{\pi-2\alpha}{4}}\,\mathrm{e}^{-\mathrm{i}\pi\left[2ut\csc\alpha-(u^2+t^2)\cot\alpha\right]}. \tag{4.18}$$

If $\alpha=\pi/2$, this kernel becomes $K_{\alpha=\pi/2}(t,u)=\mathrm{e}^{-\mathrm{i}2\pi ut}$, then $u=f$, and the fractional Fourier transform $\mathcal{F}_\alpha[x(t)]$ becomes the ordinary Fourier transform, $\hat{x}_{\alpha=\pi/2}(f)=\mathcal{F}[x(t)]$.

Given the kernel $K_\alpha(t,u)$ in Eq. (4.18), the fractional Fourier transform pair, $x(t)\leftrightarrow\hat{x}_\alpha(u)$, can be verified as follows:

$$\begin{aligned}
\mathcal{F}_\alpha^{-1}\left[\hat{x}_\alpha(u)\right] &= \int_{-\infty}^{\infty}\left(\int_{-\infty}^{\infty} x(\tau)K_\alpha(\tau,u)\mathrm{d}\tau\right)K_\alpha^*(t,u)\mathrm{d}u \\
&= \int_{-\infty}^{\infty} x(\tau)\mathrm{d}\tau\int_{-\infty}^{\infty} K_\alpha(\tau,u)K_\alpha^*(t,u)\mathrm{d}u \\
&= \int_{-\infty}^{\infty} x(\tau)\mathrm{e}^{-\mathrm{i}\pi(t^2-\tau^2)\cot\alpha}\mathrm{d}\tau\int_{-\infty}^{\infty} \mathrm{e}^{\mathrm{i}2\pi u(t-\tau)\csc\alpha}\mathrm{d}(u\csc\alpha) \\
&= \int_{-\infty}^{\infty} x(\tau)\mathrm{e}^{-\mathrm{i}\pi(t^2-\tau^2)\cot\alpha}\mathrm{d}\tau\int_{-\infty}^{\infty} \mathrm{e}^{\mathrm{i}2\pi f(t-\tau)}\mathrm{d}f \\
&= \int_{-\infty}^{\infty} x(\tau)\mathrm{e}^{-\mathrm{i}\pi(t^2-\tau^2)\cot\alpha}\,\delta(t-\tau)\mathrm{d}\tau = x(t).
\end{aligned} \tag{4.19}$$

Parseval's theorem states that energy is conserved in the ordinary Fourier transform (Eq. 1.21). In the fractional Fourier transform, $x(t)\leftrightarrow\hat{x}_\alpha(u)$, Parseval's theorem can be expressed as

$$\int_{-\infty}^{\infty}|x(t)|^2\mathrm{d}t = \int_{-\infty}^{\infty}|\hat{x}_\alpha(u)|^2\mathrm{d}u. \tag{4.20}$$

This energy conservation rule can be verified by

$$\begin{aligned}
\int_{-\infty}^{\infty}|x(t)|^2\mathrm{d}t &= \int_{-\infty}^{\infty} x^*(t)\mathrm{d}t\int_{-\infty}^{\infty}\hat{x}_\alpha(u)K_\alpha^*(t,u)\mathrm{d}u \\
&= \int_{-\infty}^{\infty}\left(\int_{-\infty}^{\infty} x(t)K_\alpha(t,u)\mathrm{d}t\right)^*\hat{x}_\alpha(u)\mathrm{d}u = \int_{-\infty}^{\infty}|\hat{x}_\alpha(u)|^2\mathrm{d}u.
\end{aligned} \tag{4.21}$$

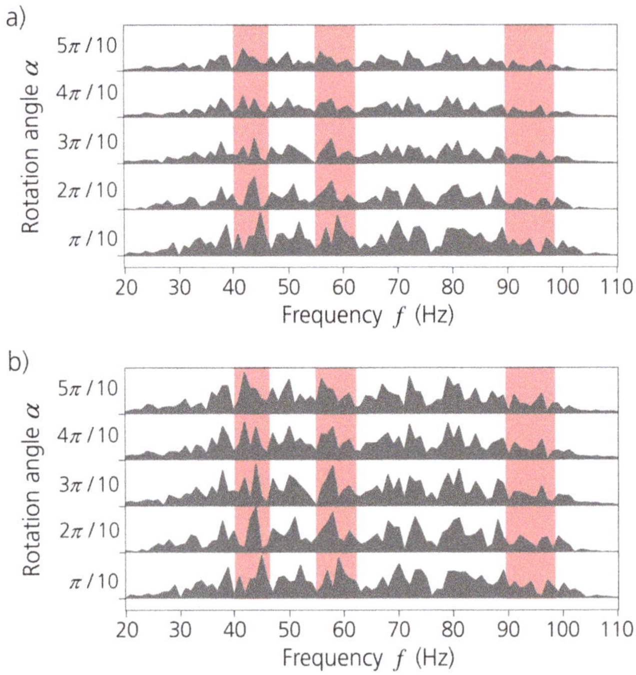

Figure 4.4 The fractional Fourier transforms. (a) The amplitude spectra produced by the fractional Fourier transforms $\mathcal{F}_\alpha[x(t)]$ with the angles of rotation $\alpha = \{5,4,3,2,1\}\times\pi/10$, respectively. (b) The normalised amplitude spectra, generated by multiplication with the factor $\sqrt{\sin\alpha}$. The horizontal axis in both subfigures is the frequency $f = u\csc\alpha$, where u is the auxiliary frequency of the fractional Fourier transform. If the angle of rotation is smaller, the amplitude spectrum $|\hat{x}_\alpha(f)|$ can separate (highlight) some energy distributions that cannot be identified in the ordinary Fourier transform (with $\alpha = \pi/2$).

The fractional Fourier transform, $x(t) \leftrightarrow \hat{x}_\alpha(u)$, can be calculated efficiently with the assistance of the ordinary Fourier transform, $x(t) \leftrightarrow \hat{x}(f)$, as summarised in Appendix D.

Figure 4.4a displays the amplitude spectra of the fractional Fourier transforms, using the sample field seismic trace that was shown in Figures 4.1 and 4.2. Note that the horizontal axis is the frequency, $f = u\csc\alpha$.

When a spectrum $\hat{x}_\alpha(f)$ is converted from the auxiliary frequency u to the ordinary frequency f, the form of energy conservation becomes

$$\int_{-\infty}^{\infty} |x(t)|^2 \mathrm{d}t = \int_{-\infty}^{\infty} \left|\sqrt{\sin\alpha}\,\hat{x}_\alpha(f)\right|^2 \mathrm{d}f. \tag{4.22}$$

Thus, the factor $\sqrt{\csc\alpha}$ in the fractional Fourier transform kernel $K_\alpha(t,u)$ is for ensuring energy conservation of the spectrum $\hat{x}_\alpha(u)$ in Eq. (4.20).

After converting the auxiliary frequency u to the ordinary frequency f, the amplitude spectrum $|\hat{x}_\alpha(f)|$ can also be normalised by multiplying the factor of $\sqrt{\sin\alpha}$, according to Eq. (4.22), for the sake of visualisation, as demonstrated in Figure 4.4b.

The five amplitude spectra in Figure 4.4 correspond to the rotation angles $\alpha = \{5,4,3,2,1\}\times\pi/10$, respectively. The top panel in each subfigure is the amplitude spectrum of the ordinary Fourier transform with $\alpha = \pi/2$. When the rotation angle α is smaller, the spectrum can separate some energy distributions, for instance, the marked zones, that cannot be identified in the ordinary Fourier transform.

4.4 The Fractional S Transform

In the S transform, if the ordinary Fourier transform kernel $e^{-i2\pi ft}$ is replaced with the fractional Fourier transform kernel $K_\alpha(t,u)$, the resultant variant of the S transform is referred to as the fractional S transform (Xu & Guo, 2012; Singh, 2015; Du *et al.*, 2016; Ranjan *et al.*, 2020).

For the definition of the fractional S transform, I propose to construct the complex signal, at any given auxiliary frequency u, as

$$y(t,u;\alpha) = x(t)R_\alpha(t,u), \tag{4.23}$$

where $R_\alpha(t,u)$ is a normalised fractional Fourier transform kernel:

$$R_\alpha(t,u) = e^{-i\frac{\pi-2\alpha}{4}} e^{-i\pi\left[2ut\csc\alpha-(u^2+t^2)\cot\alpha\right]}. \tag{4.24}$$

The normalisation is inspired by the balanced amplitude spectra, that are shown in Figure 4.4b. In the context of the fractional S transform, the phase angle is rotated within the range $0 < \alpha \le \pi/2$.

Recall that the frequency dependency in the Gaussian window function was for retaining the number of signal oscillations within a window. Therefore, for the complex signal $y(t,u;\alpha)$, corresponding to the given auxiliary frequency u, the Gaussian window function can be expressed as

$$w_p(t,u) = \sqrt{\frac{|u|^p}{k\sqrt{\pi}}}\, e^{-\frac{|u|^{2p}t^2}{2k^2}}. \tag{4.25}$$

This window function is identical to Eq. (4.13), which is used in the generalised S transform, but the frequency parameter is different. Here, it is the auxiliary frequency u.

In a similar way to the generalised S transform, the fractional S transform (FST) may also be defined as a time-domain convolution,

$$S_{\text{FST}}(t, u; \alpha) = \int_{-\infty}^{\infty} x(\tau) R_\alpha(\tau, u) w_p(t - \tau, u) \mathrm{d}\tau$$
$$= x(t) R_\alpha(t, u) * w_p(t, u), \tag{4.26}$$

where $*$ denotes convolution with respect to time t.

Converting the auxiliary frequency u to the ordinary frequency f produces the time–frequency spectrum as

$$S_{\text{FST}}(t, u; \alpha) \to S_{\text{FST}}(t, f = u \csc\alpha; \alpha). \tag{4.27}$$

By presenting the spectrum with the ordinary frequency (in Hz), the final time–frequency spectrum is directly comparable to the spectra that are generated by other variants of the S transform.

In the definition of the fractional S transform, the kernel $R_\alpha(t, u)$ equals to the fractional Fourier transform kernel $K_\alpha(t, u)$ multiplied by $\sqrt{\sin\alpha}$. I design this kernel for the S transform based on two rationales. First, the modulus of kernel $R_\alpha(t, u)$ becomes unity, which is the same as the modulus of the ordinary Fourier transform kernel $\mathrm{e}^{-\mathrm{i}2\pi ft}$. Second, because of

$$R_\alpha(t, u) = \sqrt{\sin\alpha}\, K_\alpha(t, u), \tag{4.28}$$

at any given time t, it is equivalent to multiplying the converted spectrum $\hat{x}_\alpha(f)$ by a factor of $\sqrt{\sin\alpha}$,

$$\int_{-\infty}^{\infty} \left|\sqrt{\sin\alpha}\, \hat{x}_\alpha(f)\right|^2 \mathrm{d}f = \int_{-\infty}^{\infty} \left|\sqrt{\sin\alpha}\, \hat{x}_\alpha(u)\right|^2 \mathrm{d}(u \csc\alpha)$$
$$= \int_{-\infty}^{\infty} \left|\hat{x}_\alpha(u)\right|^2 \mathrm{d}u = \int_{-\infty}^{\infty} \left|x(t)\right|^2 \mathrm{d}t, \tag{4.29}$$

which is exactly the formula of Parseval's theorem (Eq. 4.20). This form of energy conservation indicates that the energy of the final time–frequency spectrum $S_{\text{FST}}(t, f; \alpha)$, converted in Eq. (4.27), is independent of the phase angle but is comparable to the energy of seismic signal $x(t)$.

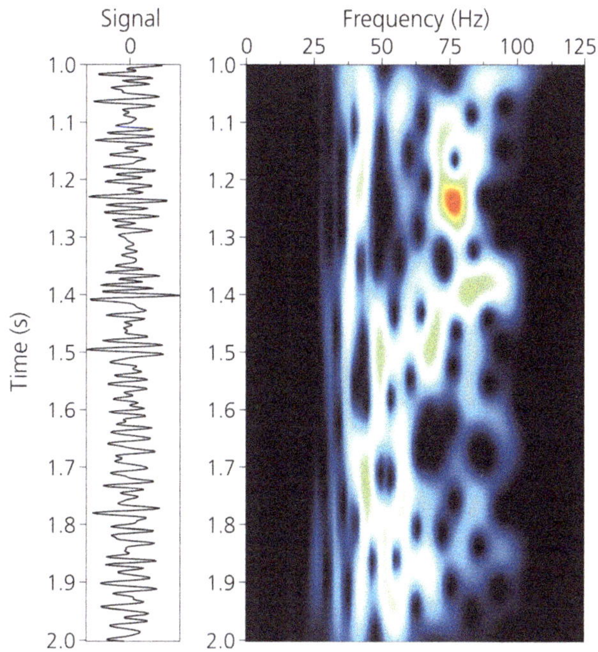

Figure 4.5 The time–frequency spectrum produced by the fractional S transform, with parameters $k = 1$, $p = 0.85$, and $\alpha = \pi/5$.

Figure 4.5 shows a time–frequency spectrum produced by the fractional S transform. The horizontal axis is the ordinary frequency f in Hz, which is converted from the auxiliary frequency parameter u via $f = u \csc \alpha$.

The fractional S transform uses the same p parameter ($p = 0.85$) as that used in the generalised S transform. However, because the fractional S transform uses the angle of rotation $\alpha = \pi/5$, it produces a time–frequency spectrum that is clearly different from the spectrum produced by the generalised S transform.

In the time–frequency spectrum produced by the fractional S transform, some spectral features can be distinguished from each other along the frequency axis, due to the angular rotation. For example, the spectrum clearly shows three distinguishable energy clouds between the times 1.35 s and 1.5 s.

How is the fractional S transform related to the fractional Fourier transform? Given the pair of the fractional Fourier transform, $x(t) \leftrightarrow \hat{x}_\alpha(u)$, the fractional S transform (Eq. 4.27) can be expressed in terms of $\hat{x}_\alpha(u)$, as

$$
\begin{aligned}
S_{\mathrm{FST}}(t,u;\alpha) &= \int_{-\infty}^{\infty} x(\tau)\left[\sqrt{\sin\alpha}\,K_\alpha(\tau,u)\right]w_p(t-\tau,u)\mathrm{d}\tau \\
&= \int_{-\infty}^{\infty}\left(\int_{-\infty}^{\infty}\hat{x}_\alpha(v)K_\alpha^*(\tau,v)\mathrm{d}v\right)\sqrt{\sin\alpha}\,K_\alpha(\tau,u)w_p(t-\tau,u)\mathrm{d}\tau \\
&= \int_{-\infty}^{\infty}\hat{x}_\alpha(v)\left(\sqrt{\sin\alpha}\int_{-\infty}^{\infty}K_\alpha(\tau,u)K_\alpha^*(\tau,v)w_p(t-\tau,u)\mathrm{d}\tau\right)\mathrm{d}v \\
&= \int_{-\infty}^{\infty}\hat{x}_\alpha(v)H_\alpha(t,u,v)\mathrm{d}v,
\end{aligned}
\tag{4.30}
$$

where $\hat{x}_\alpha(v)$ is the fractional Fourier transform, and $H_\alpha(t,u,v)$ can be derived as

$$
\begin{aligned}
H_\alpha(t,u,v) &= \sqrt{\sin\alpha}\int_{\tau=-\infty}^{\infty}K_\alpha(\tau,u)K_\alpha^*(\tau,v)w_p(t-\tau,u)\mathrm{d}\tau \\
&= \sqrt{\frac{2k\sqrt{\pi}\csc\alpha}{|u|^p}}\,\mathrm{e}^{-\frac{2k^2\pi^2(u-v)^2\csc^2\alpha}{|u|^{2p}}}\,\mathrm{e}^{\mathrm{i}\pi(u^2-v^2)\cot\alpha}\,\mathrm{e}^{-\mathrm{i}2\pi(u-v)t\csc\alpha}.
\end{aligned}
\tag{4.31}
$$

In this derivation, the Gaussian window function $w_p(t,u)$ is defined by Eq. (4.25).

4.5 Implementation of the S Transforms

The standard S transform can be understood as a time convolution between a complex signal $y(t,f)=x(t)\mathrm{e}^{-\mathrm{i}2\pi ft}$ and the Gaussian window function $w(t,f)$ (Eq. 4.8). The Fourier transform of the complex signal $y(t,f)$, at any given frequency f, can be written as

$$
\hat{y}(\xi,f)=\mathcal{F}_t\left[y(t,f)\right]=\mathcal{F}_t\left[x(t)\mathrm{e}^{-\mathrm{i}2\pi ft}\right]. \tag{4.32}
$$

The Fourier transform of the Gaussian window function $w(t,f)$ can be computed by

$$
\hat{w}(\xi,f)=\mathrm{e}^{\mathrm{i}2\pi\xi t_0}\mathcal{F}_t\left[w(t-t_0,f)\right], \tag{4.33}
$$

where t_0 is a time shift, which can be set as a half of the total length of seismic trace $x(t)$. Therefore, the standard S transform can be implemented through the forward and inverse Fourier transforms:

$$S_{\mathrm{ST}}(t,f)=\mathcal{F}_{\xi}^{-1}\left[\hat{y}(\xi,f)\hat{w}(\xi,f)\right], \tag{4.34}$$

where $\mathcal{F}_{\xi}^{-1}$ is the inverse Fourier transform with respect to frequency ξ.

The computation of the generalised S transform is identical to the implementation of the standard S transform, and can be written as

$$S_{\mathrm{GST}}(t,f;p)=\mathcal{F}_{\xi}^{-1}\left[\hat{y}(\xi,f)\hat{w}_p(\xi,f)\right]. \tag{4.35}$$

For the fractional S transform, the time convolution may also be implemented by using the Fourier transforms,

$$S_{\mathrm{FST}}(t,u;\alpha)=\mathcal{F}_{\xi}^{-1}\left[\hat{y}(\xi,u;\alpha)\hat{w}_p(\xi,u)\right], \tag{4.36}$$

where $\hat{y}(\xi,u;\alpha)=\mathcal{F}_t[y(t,u;\alpha)]$ and $\hat{w}_p(\xi,u)=\mathcal{F}_t[w_p(t,u)]$ are the ordinary Fourier transforms with respect to time t. Therefore, the inverse Fourier transform $\mathcal{F}_{\xi}^{-1}[\hat{y}(\xi,u;\alpha)\hat{w}_p(\xi,u)]$ is the ordinary inverse transform with respect to the ordinary frequency ξ. None of these Fourier transforms in Eq. (4.36) are the fractional Fourier transform.

Recall that there are frequency singularities in the S transforms, such as $|f|^{-1}$ in Eq. (4.10), $|f|^{-p}$ in Eq. (4.15), and $|u|^{-p}$ in Eq. (4.31). Because of the frequency singularity, energy conservation cannot be guaranteed near the boundary of the time–frequency spectrum $S_{\mathrm{ST}}(t,f)$ when $f\to 0$. This will lead to difficulties in recovering the discrete signal through an inverse S transform. Therefore, in practical calculations, the complex time–frequency spectrum needs to be scaled by a factor,

$$r=\left(\frac{2\int_{f=0}^{f_{\mathrm{Nyq}}}\int_{t=-\infty}^{\infty}\left|S_{\mathrm{ST}}(t,f)\right|^2\mathrm{d}f\mathrm{d}t}{\int_{-\infty}^{\infty}|x(t)|^2\mathrm{d}t}\right)^{1/2}, \tag{4.37}$$

where f_{Nyq} is the Nyquist frequency of the discrete calculation. The scaling $1/r$ is only multiplied to the amplitude spectrum $|S_{\mathrm{ST}}(t,f)|$, and after scaling, the complex spectrum $S_{\mathrm{ST}}(t,f)$ fulfils the condition of energy conservation.

After appropriate scaling, the Fourier transform-based S transforms in Eqs. (4.34–4.36) will produce identical time–frequency spectra to those generated by the time-domain implementation.

4.6 The Inverse S Transforms

The seismic trace $x(t)$ is a nonstationary signal, and the S transforms are the representations of the local Fourier spectrum at any time position t. Thus, the inverse S transforms shall reconstruct the original signal $x(t)$ at every time position separately.

The basic procedure of an inverse S transform, at any given time t, includes three steps (Schimmel & Gallart, 2005):

(1) Manipulating the S transform spectrum $S(t,f) \rightarrow \tilde{S}(t,f)$ over frequency f, to cancel the frequency-dependent amplitude from the Gaussian window function.
(2) Transforming the manipulated time–frequency spectrum $\tilde{S}(t,f)$ directly back to the time–time domain $\tilde{s}(t,\tau)$.
(3) Reconstructing the original seismic trace $x(t)$ by taking the signal value at $\tau = t$, that is, $\tilde{s}(t,\tau = t) \rightarrow x(t)$.

For the inversion of the generalised S transform, the first step is to manipulate the time–frequency spectrum $S_{\text{GST}}(t,f;p)$:

$$\tilde{S}_{\text{GST}}(t,f;p) = \sqrt{\frac{k\sqrt{\pi}}{|f|^p}}\, S_{\text{GST}}(t,f;p). \tag{4.38}$$

After this frequency-axis manipulation, the peak amplitude of the Gaussian window function becomes unity.

The second step is to perform an inverse Fourier transform of the manipulated spectrum $\tilde{S}_{\text{GST}}(t,f;p)$, with respect to frequency f:

$$\tilde{s}(t,\tau) = \int_{-\infty}^{\infty} \tilde{S}_{\text{GST}}(t,f;p) e^{i2\pi f\tau} df. \tag{4.39}$$

Recall that t is the time axis of the time–frequency spectrum, and τ is the time variable of the seismic segment.

For the third step, let us rewrite Eq. (4.38) explicitly as the following Fourier transform:

$$\tilde{S}_{\mathrm{GST}}(t,f;p)=\int_{-\infty}^{\infty}x(\tau)\mathrm{e}^{-\frac{|f|^{2p}(t-\tau)^2}{2k^2}}\mathrm{e}^{-\mathrm{i}2\pi f\tau}\mathrm{d}\tau, \tag{4.40}$$

in which the Gaussian window function is unity at $\tau=t$. Therefore, in the time–time representation $\tilde{s}(t,\tau)$, the 'diagonal' values at $\tau=t$ form the original seismic trace,

$$x(t)=\rho\tilde{s}(t,\tau=t), \tag{4.41}$$

where ρ is a scaling factor for energy conservation. If the original signal has a finite duration T, then $\rho=T$.

Integrating the three steps above, the inversion of the generalised S transform can be expressed in a single form as

$$x(t)=2\rho\int_{0}^{\infty}\sqrt{\frac{k\sqrt{\pi}}{|f|^p}}S_{\mathrm{GST}}(t,f;p)\mathrm{e}^{\mathrm{i}2\pi ft}\,\mathrm{d}f. \tag{4.42}$$

When $p=1$, it is the inversion of the standard S transform.

For the inversion of the fractional S transform, the frequency f in the time–frequency spectrum $S_{\mathrm{FST}}(t,f;\alpha)$ needs to be converted back to the auxiliary frequency u, as

$$S_{\mathrm{FST}}(t,f;\alpha)\rightarrow S_{\mathrm{FST}}(t,u{=}f\sin\alpha;\alpha). \tag{4.43}$$

Then, at any given time t, the inversion procedure follows similar inversion steps to the generalised S transform.

The first step is the frequency-dependent manipulation of the fractional S transform $S_{\mathrm{FST}}(t,u;\alpha)$:

$$\tilde{S}_{\mathrm{FST}}(t,u;\alpha)=\sqrt{\frac{k\sqrt{\pi}}{|u|^p}}S_{\mathrm{FST}}(t,u;\alpha). \tag{4.44}$$

The second step is to obtain the time–time domain signal by using an inverse transform:

$$\tilde{s}(t,\tau)=\int_{-\infty}^{\infty}\tilde{S}_{\mathrm{FST}}(t,u;\alpha)R_{\alpha}^{*}(\tau,u)\mathrm{d}u, \tag{4.45}$$

where $R_{\alpha}^{*}(\tau,u)$ is the complex conjugate of $R_{\alpha}(\tau,u)$. This inverse transform, using the kernel $R_{\alpha}^{*}(\tau,u)$, is a scaled version of the inverse fractional Fourier transform, in which the kernel is $K_{\alpha}^{*}(\tau,u)$.

The inverse transform of Eq. (4.45) for $\tilde{S}_{\mathrm{FST}}(t,u;\alpha) \to \tilde{s}(t,\tau)$, at a given time t, can be numerically calculated through the ordinary fast Fourier transform (FFT):

$$\begin{aligned}
\tilde{s}(t,\tau) &= \mathrm{e}^{\mathrm{i}\frac{\pi-2\alpha}{4}} \int_{-\infty}^{\infty} \tilde{S}_{\mathrm{FST}}(t,u;\alpha)\mathrm{e}^{\mathrm{i}\pi\left[2u\tau\csc\alpha-(u^2+\tau^2)\cot\alpha\right]}\mathrm{d}u \\
&= \mathrm{e}^{\mathrm{i}\frac{\pi-2\alpha}{4}} \mathrm{e}^{-\mathrm{i}\pi\tau^2\cot\alpha} \int_{-\infty}^{\infty} \left[\tilde{S}_{\mathrm{FST}}(t,u;\alpha)\mathrm{e}^{-\mathrm{i}\pi u^2\cot\alpha}\right]\mathrm{e}^{\mathrm{i}2\pi u\tau\csc\alpha}\mathrm{d}u \\
&= \sin\alpha\, \mathrm{e}^{\mathrm{i}\frac{\pi-2\alpha}{4}} \mathrm{e}^{-\mathrm{i}\pi\tau^2\cot\alpha} \int_{-\infty}^{\infty} \left[\tilde{Y}_\alpha(t,u)\right]\mathrm{e}^{\mathrm{i}2\pi(u\csc\alpha)\tau}\mathrm{d}(u\csc\alpha) \\
&= \sin\alpha\, \mathrm{e}^{\mathrm{i}\frac{\pi-2\alpha}{4}} \mathrm{e}^{-\mathrm{i}\pi\tau^2\cot\alpha}\mathcal{F}_f^{-1}\left[\tilde{Y}_\alpha(t,f)\right].
\end{aligned} \tag{4.46}$$

The implementation consists of the following five steps:

(1) Multiplying the fractional S transform by a chirp function: $\tilde{Y}_\alpha(t,u) = \tilde{S}_{\mathrm{FST}}(t,u;\alpha)\mathrm{e}^{-\mathrm{i}\pi u^2\cot\alpha}$.

(2) Converting the frequency coordinate of the spectrum from f to u: $\tilde{Y}_\alpha(t,u) \to \tilde{Y}_\alpha(t,f=u\csc\alpha)$.

(3) Performing an ordinary inverse Fourier transform: $\mathcal{F}_f^{-1}[\tilde{Y}_\alpha(t,f)] = \tilde{y}_\alpha(t,\tau)$.

(4) Multiplying by another chirp function: $\tilde{z}_\alpha(t,\tau) = \tilde{y}_\alpha(t,\tau)\mathrm{e}^{-\mathrm{i}\pi\tau^2\cot\alpha}$.

(5) Multiplying by a complex factor: $\tilde{s}(t,\tau) = \sin\alpha\,\mathrm{e}^{\mathrm{i}(\pi-2\alpha)/4}\,\tilde{z}_\alpha(t,\tau)$.

Because the manipulated spectrum $\tilde{S}_{\mathrm{FST}}(t,u;\alpha)$ is equivalent to the following fractional Fourier transform:

$$\tilde{S}_{\mathrm{FST}}(t,u;\alpha) = \int_{-\infty}^{\infty} x(\tau)R_\alpha(\tau,u)\mathrm{e}^{-\frac{|u|^{2p}(t-\tau)^2}{2k^2}}\mathrm{d}\tau, \tag{4.47}$$

in which the window function is unity at $\tau = t$, the original seismic signal is reconstructed by retrieving $x(t) = \rho\tilde{s}(t,\tau=t)$, including the factor $\rho = T$ for energy conservation.

An alternative approach to the inverse S transforms is to integrate the time–frequency spectrum over time. But this conventional inversion method, according to the published literature, is an approximation.

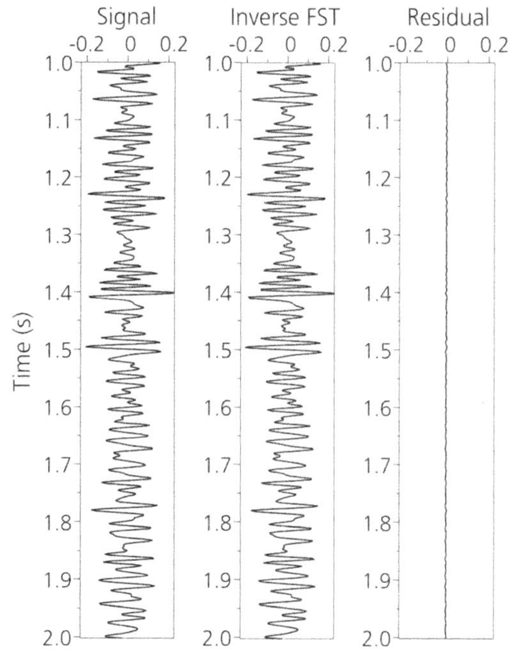

Figure 4.6 The inverse fractional S transform. (a) The original seismic trace. (b) The seismic trace reconstructed using the inverse fractional S transform. (c) The difference between the two traces.

For the inversion of the generalised S transform, one can revisit Eq. (4.15), the time average of the time–frequency spectrum, and formulate the inversion as the following double integrals:

$$x(\tau) = \int_{-\infty}^{\infty} e^{i2\pi f\tau} df \int_{-\infty}^{\infty} \sqrt{\frac{|f|^p}{2k\sqrt{\pi}}} S_{GST}(t, f; p) dt$$
$$= \int_{-\infty}^{\infty} \hat{x}(f) e^{i2\pi f\tau} df. \tag{4.48}$$

For the inverse fractional S transform, one forms the time average as

$$\hat{x}_\alpha(u) = \int_{-\infty}^{\infty} h(t, u) S_{FST}(t, u; \alpha) dt, \tag{4.49}$$

where $h(t, u)$ is an invertibility operator. Substitution of the fractional S transform Eq. (4.26) into Eq. (4.49) leads to

$$\hat{x}_\alpha(u) = \int_{-\infty}^{\infty} x(\tau) R_\alpha(\tau, u) \mathrm{d}\tau \int_{-\infty}^{\infty} h(t, u) w_p(t - \tau, u) \mathrm{d}t$$

$$= \int_{-\infty}^{\infty} x(\tau) K_\alpha(\tau, u) \mathrm{d}\tau \left(\sqrt{\sin\alpha} \int_{-T}^{T} h(t, u) w_p(t - \tau, u) \mathrm{d}t \right), \tag{4.50}$$

where $R_\alpha(\tau, u) = \sqrt{\sin\alpha}\, K_\alpha(\tau, u)$, and T is the (half-)length of the window function. It requires the following invertibility condition:

$$\sqrt{\sin\alpha} \int_{-T}^{T} h(t, u) w_p(t, u) \mathrm{d}t = 1. \tag{4.51}$$

This invertibility condition ensures that $\hat{x}_\alpha(u)$ is the fractional Fourier transform of $x(t)$. The invertibility operator $h(t, u)$ depends on both time and frequency, but, practically, it can be assumed to be a constant at a given frequency u, as

$$h(t, u) \approx \tilde{h}(u) = \frac{\sqrt{\csc\alpha}}{\int_{-T}^{T} w_p(t, u) \mathrm{d}t}. \tag{4.52}$$

Under this time-independent assumption, a simple averaging operation of the local spectra over time t would be equivalent to the fractional Fourier spectrum of the original signal:

$$\hat{x}_\alpha(u) = \tilde{h}(u) \int_{-\infty}^{\infty} S_{\mathrm{FST}}(t, u; \alpha) \mathrm{d}t. \tag{4.53}$$

Then, the signal is reconstructed by using the inverse fractional Fourier transform, $x(\tau) = \mathcal{F}_u^{-1}[\hat{x}_\alpha(u)]$ (Figure 4.6).

I would like to emphasise that these inversion formulae in Eqs. (4.48) and (4.53) are approximations since the time-averaging process in these two formulae ignores the nonstationary property. To understand this aspect, let us consider a general case where the time–frequency spectrum has been processed by phase correction. The phase correction is made time dependent because the frequency dispersion in the original seismic data depends on the travel time. This nonstationary property of the seismic phase cannot simply be averaged out.

4.7 Application to Clastic and Carbonate Reservoirs

Figures 4.7 and 4.8 show the application of this technology in the clastic and carbonate reservoirs. The lithology in the seismic section is mainly composed of two parts separated by a geologic unconformity: a series of low-frequency reflections with strong amplitudes at about 1.45 s. The shallow part (< 1.45 s) is the Permian and Carboniferous clastic formation. The target sand bodies are distributed within the interbedded coal seams. The deep part (≥ 1.45 s) is the Ordovician carbonate formation, where the hydrocarbon reservoirs consist mainly of fractures and cavities.

The target reservoirs in the clastic formation (< 1.45 s) are gas reservoirs, for which a significant feature is the low-frequency shadow. As shown in Figure 4.7, there is a large cloud which appears at frequency 19 Hz, and

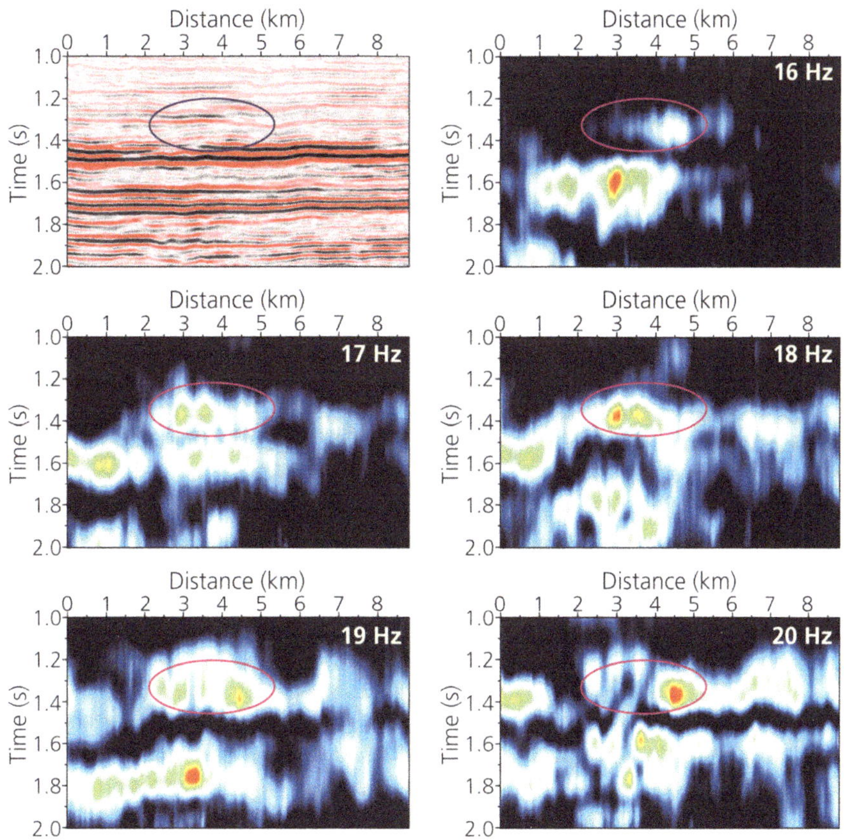

Figure 4.7 A seismic profile and the iso-frequency profiles. Gas reservoirs within a clastic formation (time < 1.45 s) produce low-frequency shadows (circled).

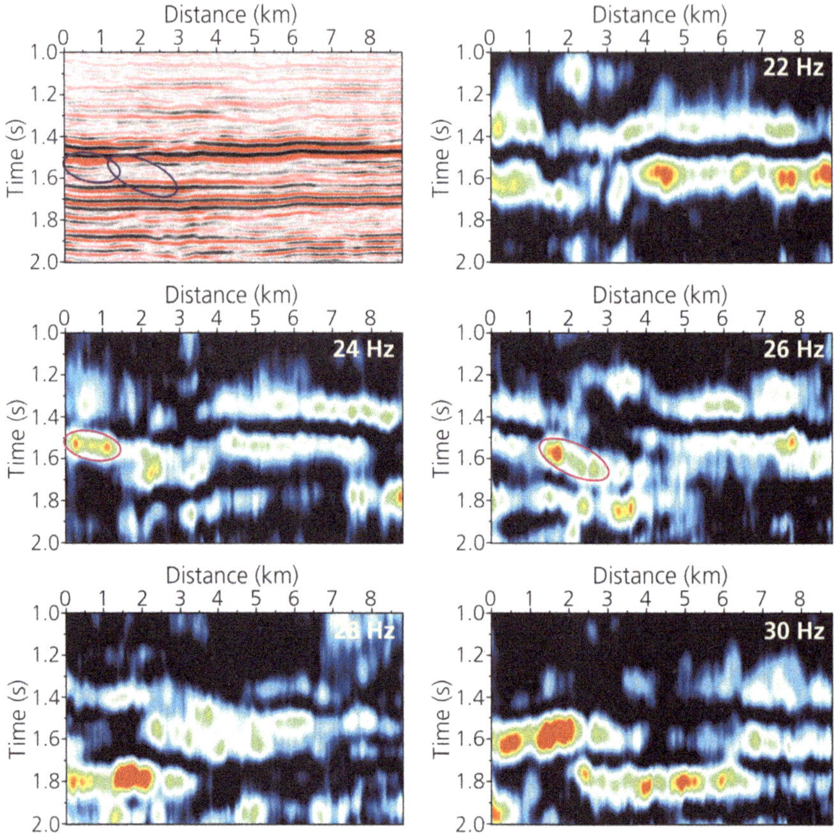

Figure 4.8 Detection of potential reservoirs within the carbonate formation (time ≥ 1.45 s). The iso-frequency profiles show the anomalies (circled) associated with fractures and cavities. These anomalies are not readily visible on the original seismic profile.

can be traced in the low-frequency spectra. But the majority of this cloud energy disappears at frequency 20 Hz.

The iso-frequency profile at 22 Hz shows that the horizontal reflections are laterally continuous. But as can be seen in Figure 4.8, the iso-frequency profiles show the location of fractures and cavities in the carbonate formation (time ≥ 1.45 s), although the original seismic profile does not readily reveal these anomalies. These fracture reservoirs do not necessarily show lateral continuity, although some locations show a dipping structure (at 26 Hz) and a horizontal structure (at 30 Hz).

5

The W Transform

The continuous wavelet transform and the S transform have a constant resolution along the frequency axis. Therefore, both methods have low temporal resolution at low frequencies and cannot adequately represent low-frequency anomalies of seismic signals. This chapter introduces the W transform, which can improve the temporal resolution of the spectrum at low frequencies. The W transform is characterised by three features:

(1) The spectral energy is localised around the dominant frequency of a seismic wavelet, instead of being shifted to higher frequencies by the continuous wavelet transform and the S transform.
(2) The Gaussian window function is nonstationary, at any frequency, along the time axis, whereas it is stationary for the continuous wavelet transform and the S transform.
(3) The implementation is numerically stable, because it avoids the potential frequency singularity that exists with the continuous wavelet transform and the S transform.

Therefore, the W transform produces a time–frequency spectrum that provides a suitable representation of the seismic properties that change with time, which is an indicator of geological depth.

5.1 Basics of the W Transform

The W transform improves the temporal resolution of the time–frequency spectrum at low frequencies and aims to effectively detect low-frequency anomalies in seismic signals (Wang, 2021).

Time–Frequency Analysis of Seismic Signals, First Edition. Yanghua Wang.

The W transform defines the Gaussian window function with a frequency weighting such that the centroid of the spectral energy distribution, at a given time t, corresponds to the dominant frequency of the wavelet at that time. The frequency weighting increases in proportion to the absolute value of the frequency difference $|\Delta f|$ from the dominant frequency. The time–frequency spectrum produced by the W transform, at any given time, therefore has symmetry at the dominant frequency.

Because of the symmetry of the frequency weighting in the Gaussian window function and thus the symmetry of the frequency spectrum, at any given time, the method is called the W transform. The resulting spectrum has a high resolution in the time axis, not only at high but also at low frequencies.

The dominant frequency of the wavelet along a seismic trace is time dependent. Therefore, both the frequency weighting and the Gaussian window function in the W transform, at any frequency, are nonstationary along the time axis.

The definition of the W transform can be started from the definition of the Gabor transform (Eq. 2.3):

$$S(t, f) = \int_{-\infty}^{\infty} x(\tau) w(t-\tau) \mathrm{e}^{-\mathrm{i}2\pi f\tau} \mathrm{d}\tau, \tag{5.1}$$

which takes a windowed signal segment and, using the Fourier transform, generates an instant frequency spectrum that is associated with the central position t of the Gaussian window function. The Gaussian window function is defined as (Eq. 2.12)

$$w(t-\tau) = \frac{1}{\sqrt[4]{\pi\sigma^2}} \mathrm{e}^{-\frac{(t-\tau)^2}{2\sigma^2}}, \tag{5.2}$$

in which the Gaussian spread parameter σ controls the breadth of the window function. In the W transform, the Gaussian spread parameter σ is defined as a time-dependent function:

$$\sigma(t, f) = \frac{k}{f_0(t) + |f - f_0(t)|}, \tag{5.3}$$

where $f_0(t)$ is the dominant frequency of the seismic signal, varying with time t. The Gaussian window function becomes

$$w(t-\tau, f) = \sqrt{\frac{f_0(t) + |f - f_0(t)|}{k\sqrt{\pi}}} \mathrm{e}^{-\frac{(t-\tau)^2}{2k^2}[f_0(t)+|f-f_0(t)|]^2}. \tag{5.4}$$

Then, Eq. (5.1) can be expressed as

$$S_{\mathrm{WT}}(t, f) = \sqrt{\frac{f_0(t) + |f - f_0(t)|}{k\sqrt{\pi}}} \int_{-\infty}^{\infty} x(\tau) \mathrm{e}^{-\frac{(t-\tau)^2}{2k^2}[f_0(t)+|f-f_0(t)|]^2} \mathrm{e}^{-\mathrm{i}2\pi f\tau} \mathrm{d}\tau. \tag{5.5}$$

This is the definition of the W transform (WT).

In the W transform, the frequency-dependent window at a given time t has the highest weight at the dominant frequency $f_0(t)$, and the weight decreases gradually as the absolute difference $|\Delta f(t)| = |f - f_0(t)|$ increases. Therefore, the frequency spectrum at time t is localised around the dominant frequency $f_0(t)$.

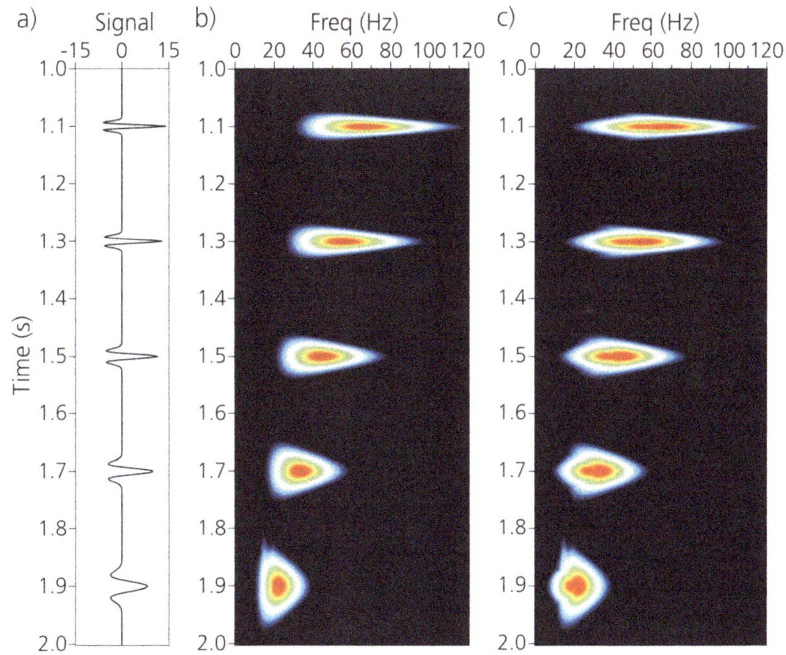

Figure 5.1 Comparison between the S transform and the W transform. (a) A synthetic seismic trace consisting of five Ricker wavelets. (b) The time–frequency spectrum produced by the S transform. (c) The time–frequency spectrum produced by the W transform, where the centre of the spectral energy distribution corresponds to the dominant frequency of the wavelet.

If we set $f_0(t) = 0$, the W transform can be deduced to the ordinary S transform. Figure 5.1 is a sample seismic trace, consisting of five Ricker wavelets, and the corresponding time–frequency spectra. The dominant frequencies of these five Ricker wavelets are $f_0 = f_p = \{60, 50, 40, 30, 20\}$ Hz, respectively.

In the spectrum that is generated by the S transform, the dominant frequency of a wavelet is at the low-frequency edge of the spectral localisation zone, and the centre of the spectral energy distribution is shifted towards a higher frequency. In contrast, in the spectrum generated by the W transform, the centre of the spectral energy distribution corresponds to the dominant frequency of a wavelet. The frequency spectrum is symmetrical around the dominant frequency.

Figure 5.2a is the same field seismic trace that is used in the previous chapters. Figure 5.2b displays the dominant frequency $f_0(t)$, varying with time t. The dominant frequency is estimated stably in terms of the weighted average of the instantaneous frequencies along the time axis (Chapter 1).

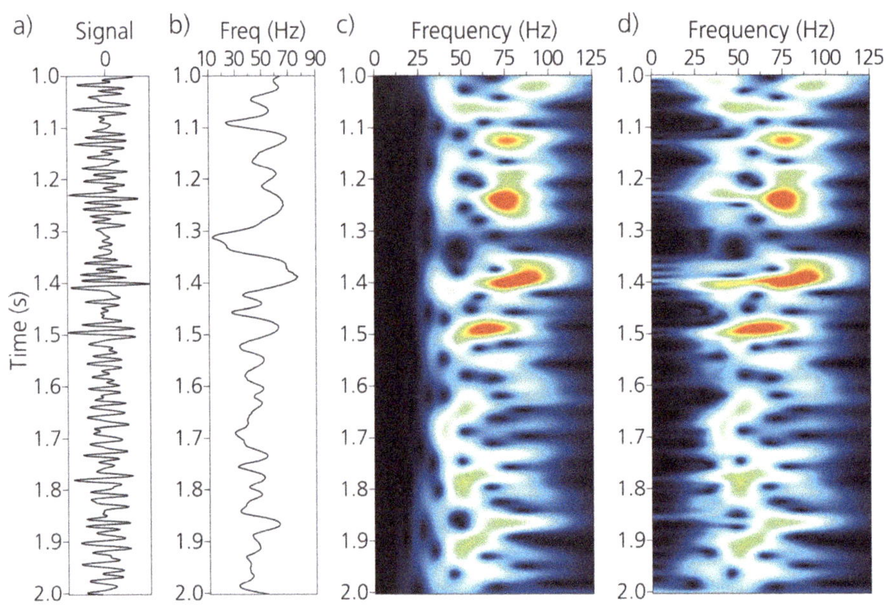

Figure 5.2 Time–frequency spectra of a field seismic trace. (a) The field seismic trace. (b) The dominant frequency, which varies with time. (c) The time–frequency spectrum produced by the S transform. (d) The time–frequency spectrum produced by the W transform.

Figures 5.2c and 5.2d compare the time–frequency spectra that are generated by the S transform and the W transform, respectively. The S transform shows a low resolution in time at low frequencies, which are the range of seismic frequencies that the reservoir detection aims at. The W transform can improve the time resolution in this low-frequency range. For example, the time–frequency spectrum reveals two energy packages at time 1.4 s. These two energy packages correspond to two waveforms, with dominant frequencies of approximately 40 Hz and 75 Hz.

According to the Gabor limit (Eq. 2.28), there is a trade-off between the improvement in time resolution and the loss of frequency resolution, for a given wavelet. Figure 5.2d clearly demonstrates that the feasible area that is covered by the time–frequency spectrum, which is proportional to the quantity $\sigma_t \sigma_f$, is relatively fixed.

The scaling factor is simply set as $k = 1$ in this example, where we focus on the analysis of other properties of the W transform.

5.2 The Generalised W Transform

While the W transform introduces a nonstationary Gaussian window function into the framework of the Gabor transform, further refinement can be made to the Gaussian window function (Li *et al.*, 2022). The W transform with a refined Gaussian window function is referred to as the generalised W transform.

The generalisation is made by modifying the spread parameter σ of the Gaussian window function in the W transform, to become a multivariate composite exponential function as

$$\sigma_p(t, f) = \frac{k}{\left[f_0^p(t) + \left| f - f_0(t) \right|^p \right]^{1/p}}, \tag{5.6}$$

where p is a trend factor that regulates the change rate of the Gaussian window with respect to the frequency variation. Then, the Gaussian window function is modified into the following form:

$$w_p(t - \tau, f) = \sqrt{\frac{\left[f_0^p(t) + \left| f - f_0(t) \right|^p \right]^{1/p}}{k\sqrt{\pi}}} \mathrm{e}^{-\frac{(t-\tau)^2}{2k^2}\left[f_0^p(t) + \left| f - f_0(t) \right|^p \right]^{2/p}}. \tag{5.7}$$

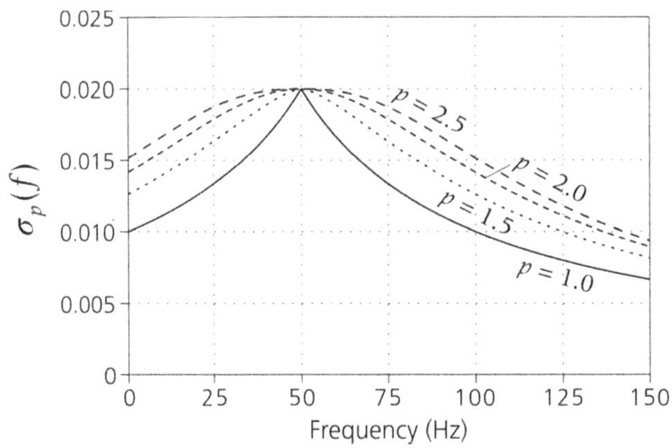

Figure 5.3 The Gaussian spread parameter $\sigma_p(f)$ versus the trend factor $p = \{1.0, 1.5, 2.0, 2.5\}$. For $p \geq 1.5$, the Gaussian spread parameter $\sigma_p(f)$ is differentiable with respect to the frequency *f*.

The generalised W transform (GWT) can be expressed as

$$S_{\mathrm{GWT}}(t, f; p) = \sqrt{\frac{\left[f_0^p(t) + \left|f - f_0(t)\right|^p\right]^{1/p}}{k\sqrt{\pi}}} \times \int_{-\infty}^{\infty} x(\tau) \mathrm{e}^{-\frac{(t-\tau)^2}{2k^2}\left[f_0^p(t)+\left|f-f_0(t)\right|^p\right]^{2/p}} \mathrm{e}^{-\mathrm{i}2\pi f\tau} \mathrm{d}\tau. \tag{5.8}$$

Recall that the scaling factor k in the W transform adjusts the breadth of the Gaussian window function. With the additional parameter, the trend factor p, the generalised W transform smooths the window function around the dominant frequency $f_0(t)$ and makes the spectral energy distribution in the time–frequency plane more localised around $f_0(t)$.

Figure 5.3 plots the generalised Gaussian spread parameter $\sigma_p(f)$, which is a function of trend factor p. Four curves are associated with $p = \{1.0, 1.5, 2.0, 2.5\}$, respectively. In this example, the scaling factor is $k = 1$, and the dominant frequency is $f_0 = 50$ Hz. When $p = 1$ (the solid curve), the Gaussian spread parameter $\sigma_{p=1}(f)$, which corresponds to that used in the ordinary W transform, clearly shows a differential singularity at the dominant frequency.

This differential singularity in the Gaussian spread parameter $\sigma_{p=1}(f)$ leads to a differential singularity in the window function. Figure 5.4 plots

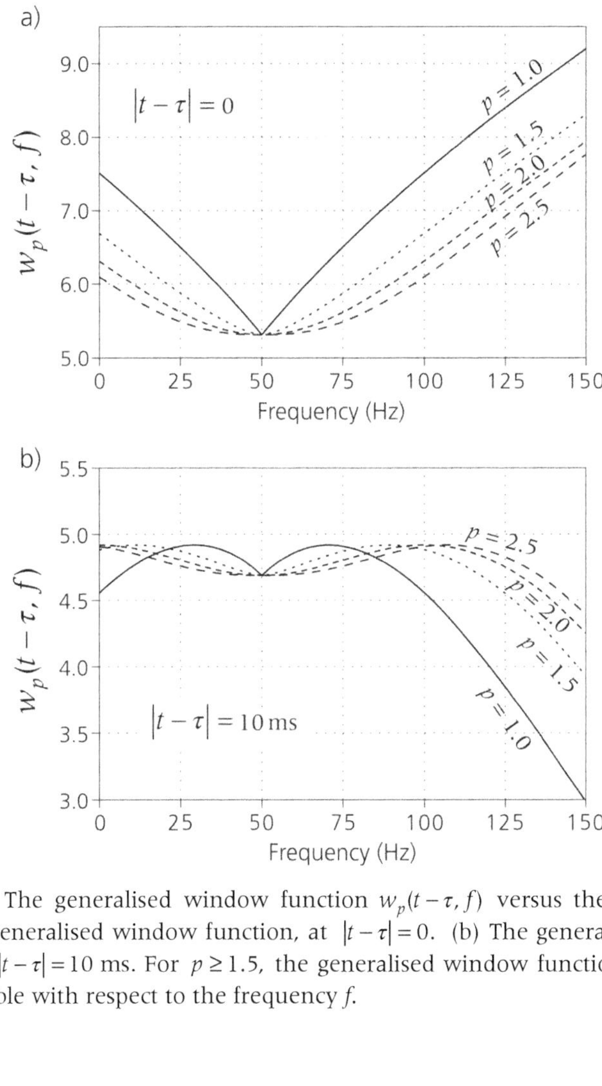

Figure 5.4 The generalised window function $w_p(t-\tau, f)$ versus the trend factor p. (a) The generalised window function, at $|t-\tau| = 0$. (b) The generalised window function, at $|t-\tau| = 10$ ms. For $p \geq 1.5$, the generalised window function $w_p(t-\tau, f)$ is differentiable with respect to the frequency f.

the generalised Gaussian window $w_p(t-\tau, f)$, at $|t-\tau| = 0$ and 10 ms. The rest of the parameters are identical to those used in Figure 5.3.

There is no doubt that the window function $w_p(t-\tau, f)$, at any given frequency f, is in the Gaussian form along the time axis. However, in the ordinary W transform with $p = 1$, the window function (the solid curves) appears to have a differential singularity with respect to the frequency, for any given time. In contrast, in the generalised W transform, for values of $p \geq 1.5$, the window function (dotted and dashed curves) is not only continuous but is also smooth with respect to the frequency.

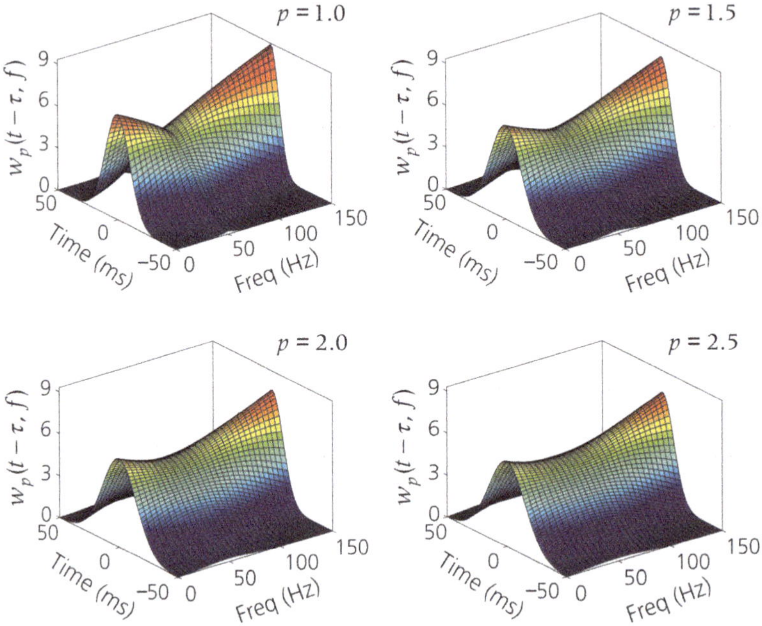

Figure 5.5 The generalised window function $w_p(t-\tau, f)$ with different trend factors, $p = \{1.0,\ 1.5,\ 2.0,\ 2.5\}$. The time axis is $t-\tau$. The generalised window function $w_p(t-\tau, f)$, for $p \geq 1.5$, is differentiable with respect to the frequency f.

When $p \to \infty$, parameter $\sigma_p(f)$ is only related to the dominant frequency $f_0(t)$:

$$\lim_{p \to \infty} \sigma_p(f) = \frac{k}{f_0(t)}, \tag{5.9}$$

if assuming $|f - f_0(t)|/f_0 < 1$. In this case, the Gaussian spread parameter σ is independent of the frequency, and the generalised W transform degenerates into a Gabor transform. However, the Gabor transform here has a time-varying window breadth, according to the time-varying dominant frequency $f_0(t)$. Therefore, theoretically, the trend factor p could be set as $1 \leq p < \infty$.

Figure 5.5 further visualises the generalised window function $w_p(t-\tau, f)$, with different values of trend factor p. The time axis is $t-\tau$. Figures 5.4 and 5.5 reveal that the window function is differentiable with respect to the frequency, if $p \geq 1.5$. This means that the window function shares a common tangent direction, at the position of joint dominant frequency.

The smooth property here is referred to as the differentiability of the window function.

The visualisations in Figures 5.4 and 5.5 suggest that, in practice, the trend factor p can be set within the range of [1.5, 3.0], for a smooth spectral distribution along the frequency axis.

Figure 5.6 compares the W transform spectra of the synthetic trace (Figure 5.6a). Figure 5.6b is the spectrum of the ordinary W transform, when the trend factor $p = 1$. Figure 5.6c is the spectrum of the generalised W transform, when the trend factor is set as $p = 3$. The generalised W transform produces a smooth spectrum along the frequency axis.

The scalar factor k can also be adjusted to balance the time and frequency resolutions. In this example, the scalar factor k is set as $k = 0.8$ for generating both spectra.

Figure 5.7 compares the W transform spectra of the field seismic trace (Figure 5.7a). Figures 5.7b–c are the spectrum $S_{WT}(t, f; p = 1)$, produced by

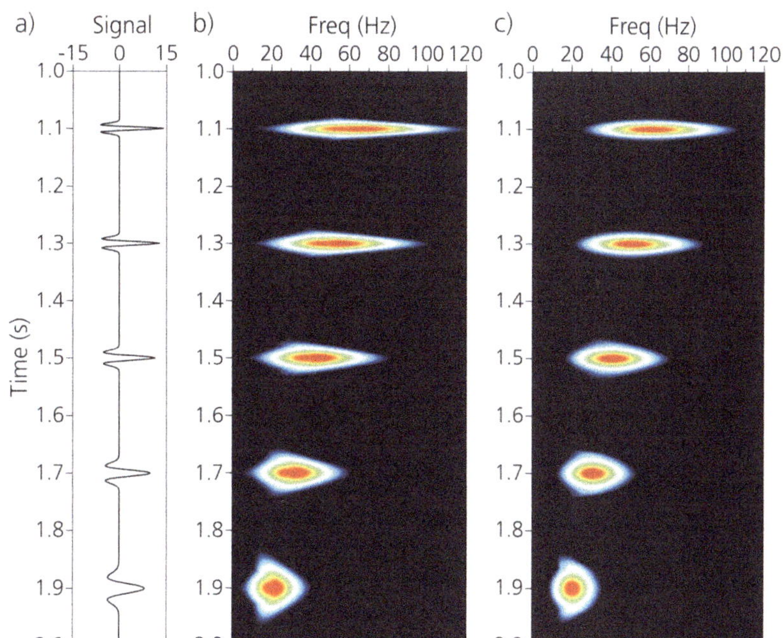

Figure 5.6 The generalised W transform with different p values. (a) The synthetic seismic trace. (b) The time–frequency spectrum produced by the ordinary W transform ($p = 1$). (c) The time–frequency spectrum produced by the generalised W transform ($p = 3$). The scaling factor is $k = 0.8$.

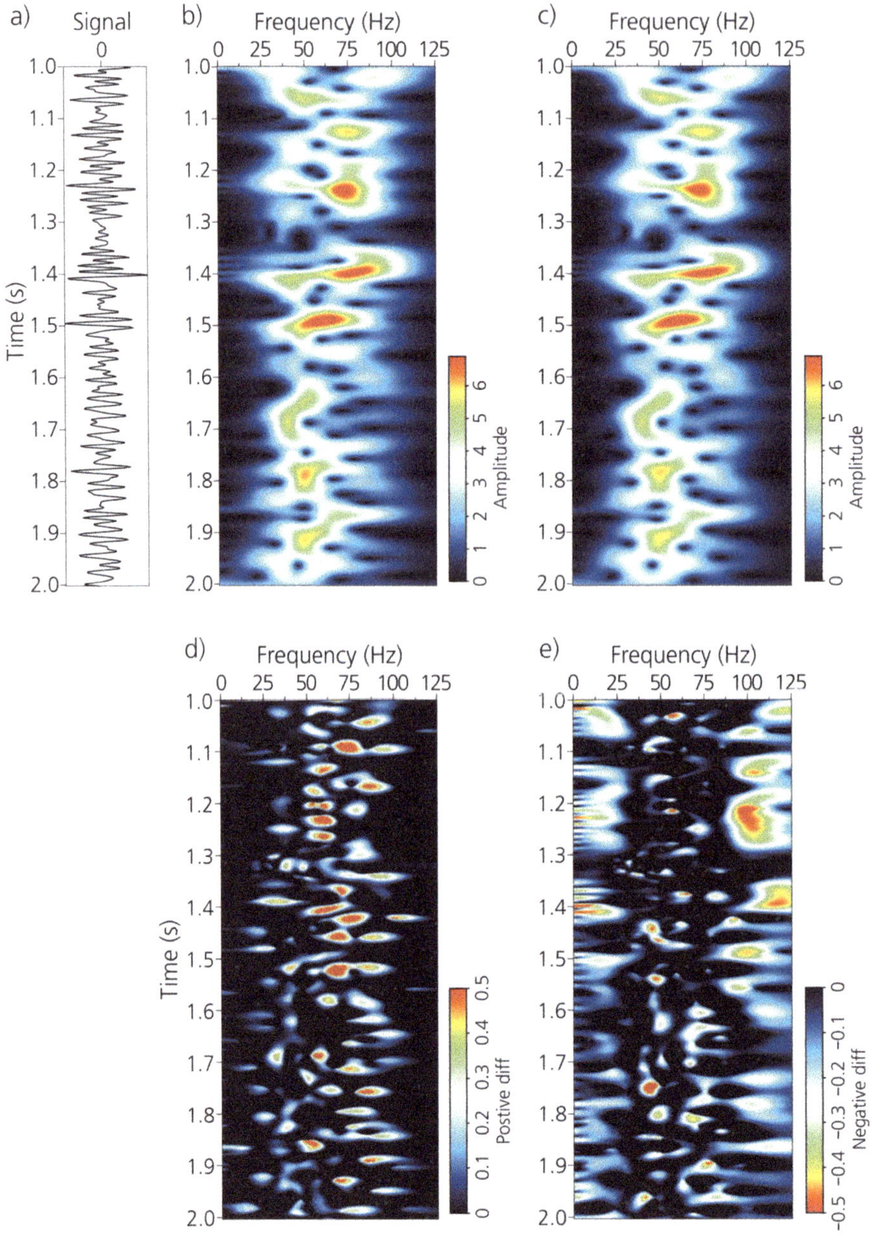

Figure 5.7 Comparison between the ordinary and the generalised W transforms. (a) The field seismic trace. (b) The time–frequency spectrum $S_{\mathrm{WT}}(t, f; p = 1)$ produced by the ordinary W transform. (c) The time–frequency spectrum $S_{\mathrm{GWT}}(t, f; p = 3)$ produced by the generalised W transform. (d, e) The positive and negative difference between the spectrum of the generalised W transform and the spectrum of the ordinary W transform.

the ordinary W transform, and the spectrum $S_{\mathrm{GWT}}(t, f; p = 3)$, produced by the generalised W transform, respectively. They use the same parameters that were used in the synthetic trace example.

To illustrate the improvement, Figure 5.7 also shows the difference between two amplitude spectra,

$$\Delta\left|S(t, f)\right| = \left|S_{\mathrm{GWT}}(t, f; p = 3)\right| - \left|S_{\mathrm{WT}}(t, f; p = 1)\right|. \tag{5.10}$$

Figure 5.7d displays the positive difference values, $\Delta|S(t, f)| \geq 0$. It clearly shows that the generalised W transform has enhanced the peaks (bright colour spots) of the amplitude distributions.

Figure 5.7e displays the negative difference values, $\Delta|S(t, f)| \leq 0$. The bright colour distributions in the ultralow-frequency area and the high-frequency area are the amplitudes that are presented in the ordinary W transform spectrum, and they have been smoothed out by the generalised W transform.

However, the improvement that is made by the generalised W transform is rather minor. The quantity of energy localisation in a time–frequency spectrum can be measured by the Rényi entropy, as in the following:

$$H_\alpha = \frac{1}{1-\alpha} \log_2 \int_{-\infty}^{\infty} \int_{-\infty}^{\infty} \left(\frac{\left|S(t, f)\right|}{\int_{-\infty}^{\infty} \int_{-\infty}^{\infty} \left|S(t, f)\right| \mathrm{d}f \mathrm{d}t} \right)^\alpha \mathrm{d}f \mathrm{d}t, \tag{5.11}$$

where $\alpha \neq 1$ is the order of the Rényi entropy (Rényi, 1961; Stanković, 2001). As indicated by the Rényi entropy curves shown in Figure 5.8, the time–frequency spectrum $S_{\mathrm{GWT}}(t, f; p = 3)$ generated by the generalised W

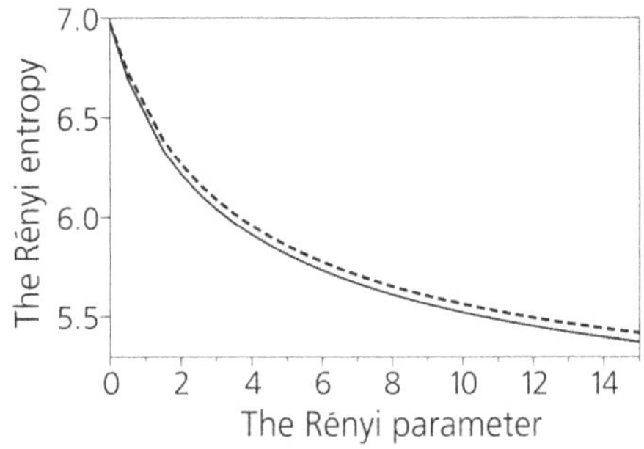

Figure 5.8 The Rényi energy of the W transform spectra. The dashed curve and the solid curve correspond to the ordinary W transform $S_{\mathrm{WT}}(t, f; p = 1)$ and the generalised W transform $S_{\mathrm{GWT}}(t, f; p = 3)$, respectively.

transform has a smaller Rényi entropy (the solid curve) than $S_{\text{WT}}(t, f; p = 1)$ generated by the ordinary W transform (the dashed curve). But the difference between the two Rényi entropies is small.

5.3 Implementation of Nonstationary Convolution

A significant feature of the W transform, in contrast to the Gabor transform, the continuous wavelet transform, and the S transform, is that the W transform uses a nonstationary Gaussian window function, varying along with time t. Therefore, the integral form of the generalised W transform can be expressed as a nonstationary time convolution:

$$\tilde{S}_{\text{GWT}}(t, \tau, f; p) = x(\tau)\mathrm{e}^{-\mathrm{i}2\pi f\tau} * w_p(t, \tau, f), \tag{5.12}$$

where $w_p(t, \tau, f) = w_p(t - \tau, f)$ is the Gaussian window function defined by Eq. (5.7). At any given frequency f, the Gaussian window function $w_p(t, \tau, f)$ depends on time t.

Equation (5.12) is a nonstationary convolution with respect to time τ, and it can be implemented through forward and inverse Fourier transforms, as

$$\tilde{S}_{\text{GWT}}(t, \tau, f; p) = \mathcal{F}_\xi^{-1}\left[\hat{y}(\xi, f)\hat{w}_p(t, \xi, f)\right], \tag{5.13}$$

where ξ is the frequency counterpart of time τ, and $\hat{y}(\xi, f)$ is the Fourier transform of the complex signal $y(\tau, f) = x(\tau)\mathrm{e}^{-\mathrm{i}2\pi f\tau}$, given by

$$\hat{y}(\xi, f) = \mathcal{F}_\tau\left[y(\tau, f)\right] = \mathcal{F}_\tau\left[x(\tau)\mathrm{e}^{-\mathrm{i}2\pi f\tau}\right]. \tag{5.14}$$

The Fourier transform of the nonstationary Gaussian window function $w_p(t, \tau, f) = w_p(t - \tau, f)$ can be derived analytically as

$$\hat{w}_p(t, \xi, f) = \mathcal{F}_\tau\left[w_p(t - \tau, f)\right] = \sqrt{\frac{2k\sqrt{\pi}}{\left[f_0^p(t) + \left|f - f_0(t)\right|^p\right]^{1/p}}}\,\mathrm{e}^{-\frac{2k^2\pi^2\xi^2}{\left[f_0^p(t)+\left|f - f_0(t)\right|^p\right]^{2/p}}}\mathrm{e}^{-\mathrm{i}2\pi\xi t}. \tag{5.15}$$

Finally, the generalised W transform spectrum $S_{\text{GWT}}(t, f; p)$ in the time–frequency domain is constructed by

$$S_{\text{GWT}}(t,f;p) = \tilde{S}_{\text{GWT}}(t,\tau = t,f;p). \tag{5.16}$$

The nonstationary property is an important property of the W transform. To understand this property, let us compare the W transform to the S transform directly. In the generalised S transform, the Gaussian window is a stationary function, and the stationary convolution (Eq. 4.14) is equivalent to a multiplication in the frequency domain (Eq. 4.35), which can be expressed explicitly as

$$S_{\text{GST}}(t,f;p) = \sqrt{2\sqrt{\pi}\sigma_p(f)} \int_{-\infty}^{+\infty} \hat{y}(\xi,f) e^{-2\pi^2\xi^2\sigma_p^2(f)} e^{i2\pi\xi t} d\xi. \tag{5.17}$$

The integrand is stationary, because at any given frequency f, the Gaussian spread parameter $\sigma_p(f) = k/|f|^p$ (Eq. 4.12) is independent from time t. Therefore, the frequency integral of Eq. (5.17) can be calculated efficiently through only a single fast implementation of the inverse Fourier transform.

But the generalised W transform is nonstationary. Equation (5.13) can be expressed as

$$\tilde{S}_{\text{GWT}}(t,\tau,f;p) = \sqrt{2\sqrt{\pi}\sigma_p(t,f)} \int_{-\infty}^{+\infty} \hat{y}(\xi,f) e^{-2\pi^2\xi^2\sigma_p^2(t,f)} e^{i2\pi\xi(\tau-t)} d\xi. \tag{5.18}$$

At any given frequency f, the Gaussian spread parameter $\sigma_p(t,f)$ (Eq. 5.6) is time dependent. Because of the nonstationary property, the frequency integral of Eq. (5.18) should be performed by a series of inverse Fourier transforms along the time sample positions t. Thus, the computation cost of the W transform is linearly proportional to the number of samples in time t rather than only a single implementation of the inverse Fourier transform in the S transform.

Numerical stability is another important feature of the W transform. In the S transform, for instance, the integrand of the frequency integral form is

$$\phi(\xi,f) = \sqrt{\frac{2k\sqrt{\pi}}{|f|^p}}\, \hat{y}(\xi,f) e^{-\frac{2k^2\pi^2\xi^2}{|f|^{2p}}}. \tag{5.19}$$

The frequency factor $|f|^{-p}$ presents a singularity issue in the numerical implementation. However, in the generalised W transform of Eq. (5.18), the integrand can be written explicitly as

$$\phi(t,\xi,f) = \sqrt{\frac{2k\sqrt{\pi}}{\left[f_0^p(t) + |f - f_0(t)|^p\right]^{1/p}}}\,\hat{y}(\xi,f)\mathrm{e}^{-\frac{2k^2\pi^2\xi^2}{\left[f_0^p(t)+|f-f_0(t)|^p\right]^{2/p}}}. \tag{5.20}$$

There is no singularity issue, because $f_0(t) > 0$ if there is any valid signal component at time t.

5.4 The Inverse W Transform

Considering the nonstationary property of the original seismic trace and the time-varying property of the frequency spectrum $S_{\mathrm{GWT}}(t,f;p)$ generated by the generalised W transform, the inverse W transform must be implemented in a time-dependent manner.

At any given time t, the implementation of the inverse W transform consists of three basic steps.

(1) Manipulating the time–frequency spectrum $S_{\mathrm{GWT}}(t,f;p)$ over frequency f, by using a frequency-dependent scaling factor.
(2) Transforming the manipulated spectrum $\tilde{S}_{\mathrm{GWT}}(t,f;p)$ directly back to the time–time domain $\tilde{s}(t,\tau)$.
(3) Reconstructing seismic trace $x(t)$ by taking the signal value at $\tau = t$.

The first step, manipulating the spectrum $S_{\mathrm{GWT}}(t,f;p)$, is to cancel the frequency-dependent amplitude of the Gaussian window function, at a given time t. The manipulated time–frequency spectrum $\tilde{S}_{\mathrm{GWT}}(t,f;p)$ is given by

$$\tilde{S}_{\mathrm{GWT}}(t,f;p) = \sqrt{\frac{k\sqrt{\pi}}{\left[f_0^p(t) + |f - f_0(t)|^p\right]^{1/p}}}\,S_{\mathrm{GWT}}(t,f;p). \tag{5.21}$$

The generalised W transform is defined in Eq. (5.8), and thus Eq. (5.21) can be expressed as a Fourier transform:

$$\tilde{S}_{\mathrm{GWT}}(t,f;p) = \int_{-\infty}^{\infty} x(\tau)\mathrm{e}^{-\frac{(t-\tau)^2}{2k^2}\left[f_0^p(t)+|f-f_0(t)|^p\right]^{2/p}}\,\mathrm{e}^{-\mathrm{i}2\pi f\tau}\mathrm{d}\tau. \tag{5.22}$$

This expression reveals that, after manipulation, the peak amplitude of the Gaussian window function becomes a unity, at $\tau = t$.

The second step is an inverse Fourier transform of the manipulated spectrum $\tilde{S}_{\mathrm{GWT}}(t, f; p)$, with respect to frequency f, as

$$\tilde{s}(t, \tau) = \int_{-\infty}^{\infty} \tilde{S}_{\mathrm{GWT}}(t, f; p) \mathrm{e}^{\mathrm{i}2\pi f \tau} \mathrm{d}f. \tag{5.23}$$

Recall that t is the time axis of the time–frequency spectrum, and τ corresponds to the time of the original seismic trace.

The third step is to formulate the original seismic trace. Comparison between Eqs. (5.22) and (5.23) indicates that the time–time domain representation $\tilde{s}(t, \tau)$ is

$$\tilde{s}(t, \tau) = x(\tau) \mathrm{e}^{-\frac{(t-\tau)^2}{2k^2}\left[f_0^p(t)+|f-f_0(t)|^p\right]^{2/p}}. \tag{5.24}$$

Therefore, the seismic trace $x(t)$ is formed by $\tilde{s}(t, \tau)$ at $\tau = t$,

$$x(t) = \rho \tilde{s}(t, \tau = t), \tag{5.25}$$

where ρ is the energy conservation factor, and $\rho = T$, the finite time duration in the numerical calculation.

5.5 Application to Detecting Hydrocarbon Reservoirs

The W transform aims to improve the localisation of the frequency spectrum and increase the temporal resolution at low frequencies. Let us demonstrate its effectiveness with an application example in detecting hydrocarbon reservoirs.

Low-frequency shadows in the time–frequency spectrum can be used as direct indicators in the detection of gas reservoirs (Castagna et al., 2003; Ebrom, 2004). One of the causes is that the high-frequency energy of seismic waves reflected from a gas-saturated reservoir is attenuated more significantly (Wang, 2010). Other causes of low-frequency amplitude anomaly are the low-frequency tuning effect of the converted slow P-wave in the gas-saturated media. However, regardless of the mechanism, the S transform may not be able to detect the low-frequency anomalies in seismic data, as clearly shown in Figure 5.2.

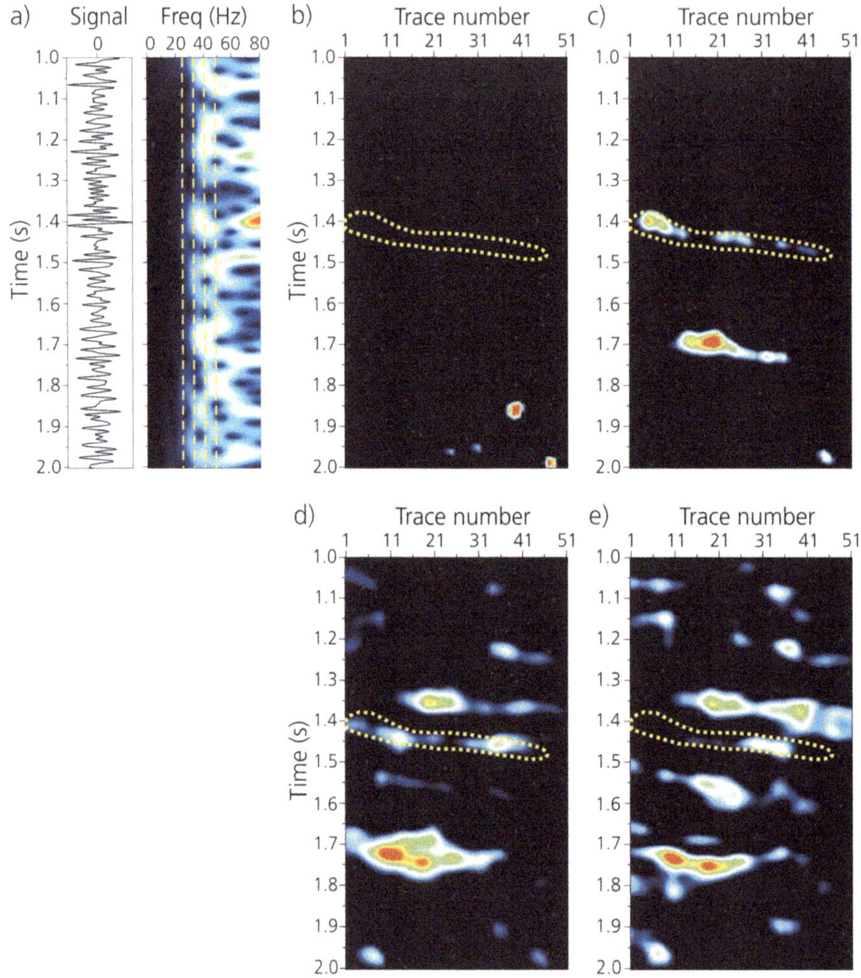

Figure 5.9 The time–frequency spectrum produced by the S transform. (a) The single seismic trace and the time–frequency spectrum produced by the S transform. (b–e) Four spectral profiles, at frequencies $f = \{24, 32, 40, 48\}$ Hz, respectively. The S transform is insensitive to the low-frequency anomaly (dotted circle).

Figure 5.9a is a replication of Figures 5.2a and 5.2c, the single seismic trace and the corresponding time–frequency spectrum that is generated by the S transform. Four vertical dashed lines mark the positions of four relatively low frequencies at $f = \{24, 32, 40, 48\}$Hz.

The field seismic profile was shown in Figure 1.6a. Figures 5.9b–e are the spectral profiles of four iso-frequencies, in which the spectral profile at 24 Hz does not provide any useful information. Because the S transform has

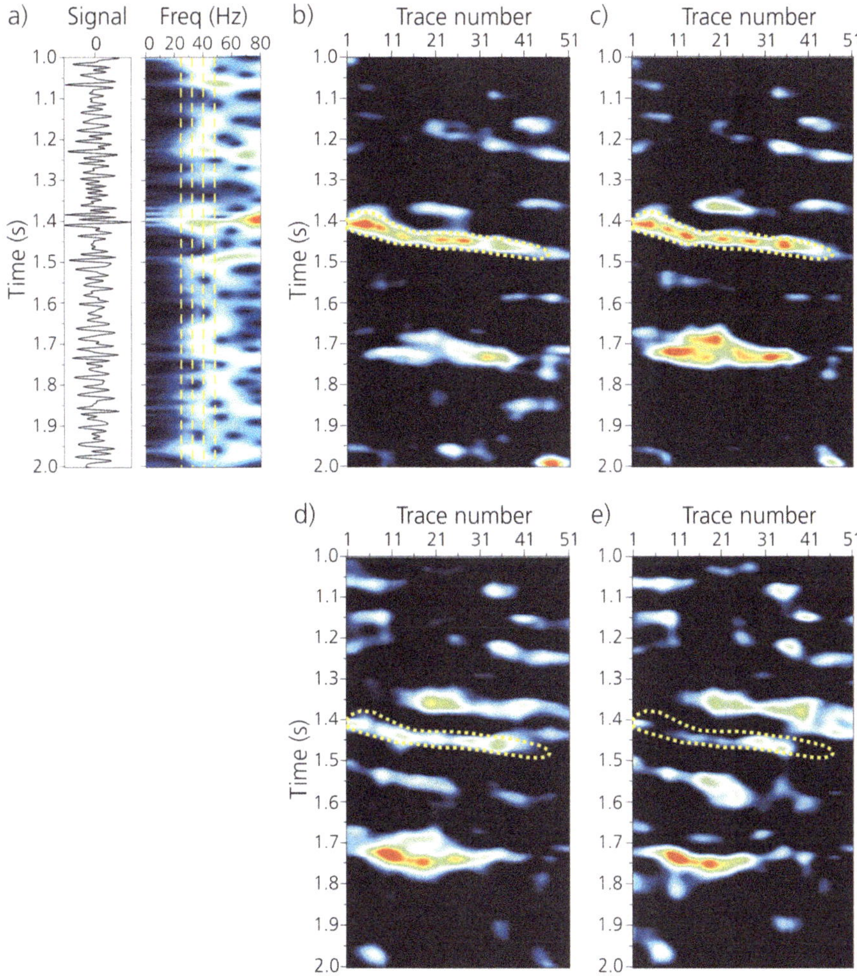

Figure 5.10 The time–frequency spectrum produced by the W transform. (a) The single seismic trace and the time–frequency spectrum produced by the W transform. (b–e) Four spectral profiles at frequencies $f = \{24, 32, 40, 48\}$ Hz, respectively. These spectral profiles show the low-frequency anomaly (dotted circle), disappearing at the medium and high frequencies.

shifted the centre of the spectral energy distribution wrongly towards high frequencies, it cannot present the low-frequency shadow that is related to a hydrocarbon reservoir (dotted circle) in a proper manner.

Comparing the W transform to the S transform, we can see that the significant difference occurs at low-frequency profiles. Figure 5.10a shows the same single seismic trace, but the time–frequency spectrum is generated by the W transform. Figures 5.10b–e are the spectral profiles of four iso-frequencies. The spectral profiles at $f = \{24, 32, 40\}$ Hz (Figures 5.10b–d) reveal a low-frequency shadow at time $t = 1.8$ s (in the first trace on the left). This is because the W transform has placed the centroid of the spectral energy distribution at the correct frequency position. The low-frequency shadow evidently disappears when the frequency is equal to or greater than 48 Hz (Figure 5.10e).

5.6 Application to Detecting Karst Voids

Here, we show another example from the field of geotechnical engineering, for validating subgrade construction (native material underneath the constructed track) of a high-speed rail track.

Detecting near-surface anomalies, such as karst cavities and voids, is an extremely important task for subgrade construction. One of the near-surface geophysics technologies is the constant-offset profiling in a seismic survey, which records seismic reflections and refractions with a constant-offset shot-receiver pair that moves along the surface. The collection of these constant-offset traces forms a seismic imaging profile.

In the example that is shown in Figure 5.11a, the constant offset of a source–receiver pair is 12 m, the spatial interval between traces is 2 m, and the time sampling of the seismic trace is 0.125 ms. In the seismic imaging profile, there are significant pull-down concave anomalies in both the first and second reflection events at a distance of around 248 m. At this position, the second event forms a funnel-shaped seismic anomaly with significant frequency decrease in the waveforms. The abnormal characteristic of the seismic waveforms indicates the development of a karst cavity (marked by an ellipse).

Figure 5.11b presents the corresponding dominant-frequency profile, estimated by the temporal average of the instantaneous frequencies (Eq. 1.40). In the temporal averaging, the low-pass filter $L(t)$ is set with the standard deviation $\sigma_t = 6\Delta t$, for seismic traces with such a dense sampling rate. In the dominant-frequency profile, there is a vertical zone with anomalously low frequency values (the dashed ellipse), which should be related to geological anomalies.

The anomalous zone is centred laterally at a distance of 248 m. The seismic trace at this lateral position, and the corresponding dominant frequency curve, are displayed in Figures 5.12a–b. The time–frequency

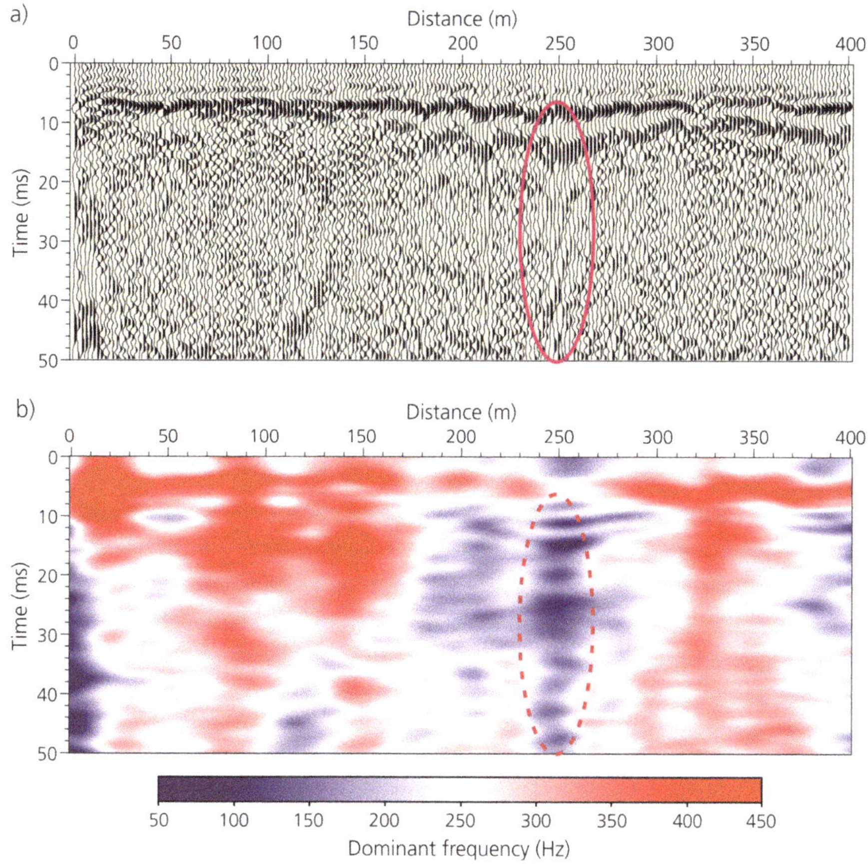

Figure 5.11 An anomalous zone (marked by the ellipse) in a seismic profile and the dominant-frequency profile. (a) The seismic constant-offset profile, in which the source–receiver offset is 12 m. (b) The dominant-frequency profile, estimated as the time average of the instantaneous frequencies.

spectrum in Figure 5.12c demonstrates again that the S transform cannot present the low-frequency anomalies properly. In contrast, the time–frequency spectrum shown in Figure 5.12d, which is generated by the W transform, clearly reveals a series of low-frequency anomalies along the vertical direction.

Figure 5.13 displays spectral profiles, that are generated by the W transform, at frequencies $f = \{80, 100, 120, 140\}$ Hz, respectively. These spectral profiles show a horizontal low-frequency zone (I) at distances between 20 m and 160 m and approximate time $t = 10$ ms. This low-frequency zone corresponds to the strong reflection event in the seismic

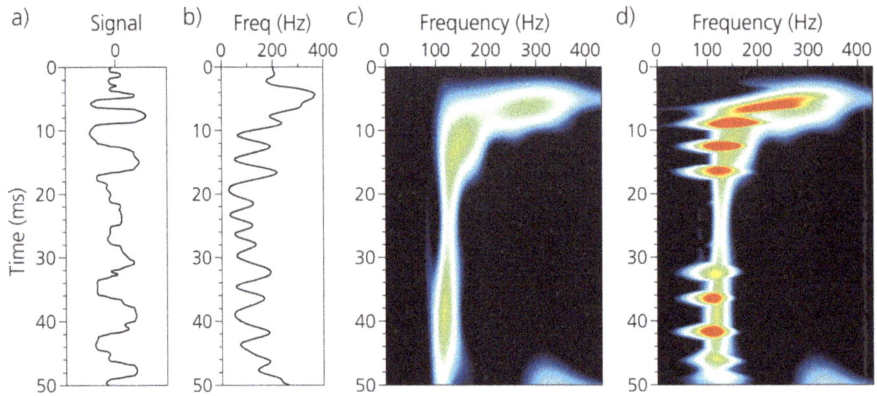

Figure 5.12 A selected seismic trace and the spectra. (a) The seismic trace at a distance of 248 m, selected from the constant-offset profile. (b) The corresponding dominant frequency curve. (c) The amplitude spectrum in the time–frequency domain produced by the S transform. (d) The amplitude spectrum in the time–frequency domain produced by the W transform.

profile. It is caused by fragmentised rocks and artificially backfilled debris during the subgrade construction. A part of these fragmentised rocks also generates seismic multiple reflections, which are the spectral anomalies (II) at distances between 105 m and 130 m and at approximate time $t = 40$ ms.

In addition, there is a 'string of beads' structure low-frequency anomaly (III) at distance 248 m. This vertical anomaly, which has the strongest amplitude, roughly at 100–120 Hz, is caused by a karst cavity above it. Subsequent drilling validated that there was a karst cavity of 1.1 m in diameter at a depth of 10.1 m (IV).

Figure 5.14 displays rock cores that were collected by drilling at distance 248 m. Cores are displayed in rows, and each row of cores is 1 m in depth. Sequentially in depth, the cores revealed: a concrete layer (0–2 m), fragmentised rocks and debris (2–2.9 m), limestone (2.9–10.1 m), a karst cave (10.1–11.2 m), and limestone (11.2–21.9 m). Cavities with different scales can be seen within the limestone segments. The mutation in spectral characteristics indicates a dramatic change in the near-surface geological conditions. In the karst cavities and caves, the severe loss of the high-frequency components of seismic waves leads to a drastic reduction in the frequency of seismic waveforms and the formation of ultralow and low-frequency anomalies, and sometimes a vertical 'string of beads' structure.

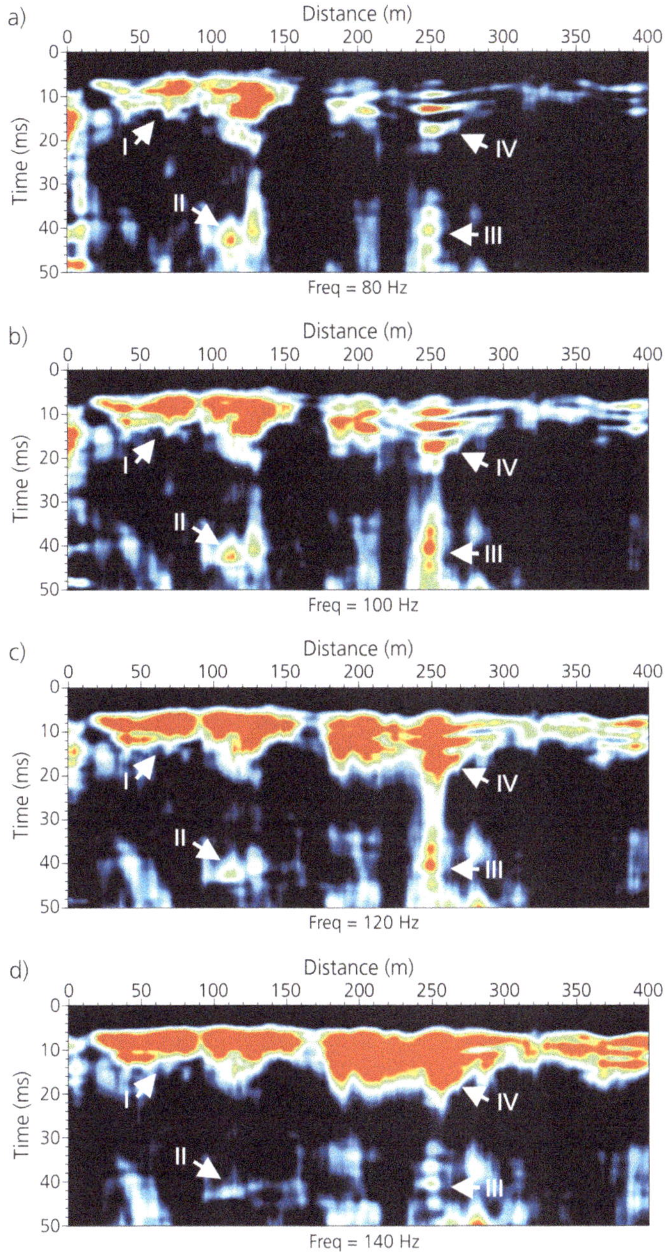

Figure 5.13 The amplitude spectra produced by the W transform. (a–d) The iso-frequency profiles at $f = \{80, 100, 120, 140\}$ Hz, respectively. The arrows indicate a horizontal low-frequency zone (I), low-frequency shadows (II), a low-frequency anomaly in the form of a 'string of beads', caused by a karst cavity above (III), and the karst cavity (IV).

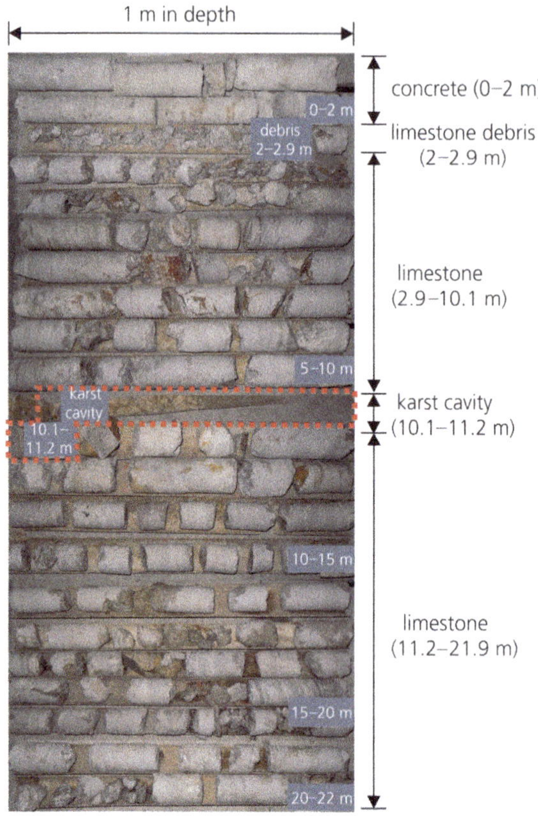

Figure 5.14 Cores reveal near-surface geological anomalies, that correspond to the low-frequency anomalies of seismic spectra. Rock cores are displayed in rows, and each row of cores is 1 m in depth.

In contrast, the rock cores at distance 110 m did not show a similar karst cave. Instead, they showed several layers of concrete mixed with backfilled rock rubble in the upper area (0–2.0 m) and limestone in the remaining area (2.0–16.2 m).

6
The Wigner–Ville Distribution

The Gabor transform-type methods presented in the first part of this book, including the Gabor transform itself, the continuous wavelet transform, the S transform, and the W transform, are *linear representations* of a nonstationary signal in the time–frequency plane. They are essentially a series of frequency spectra that change with time, and they show what spectral content is present at each time position.

To improve the resolution of linear representations, *quadratic representations* can be used in nonstationary signal analysis. There are two motivations for using quadratic representations: energetic interpretation and correlative interpretation. Energetic interpretation aims to combine temporal energy and spectral energy, while correlative interpretation aims to combine temporal correlation and spectral correlation. These combinations lead to an improvement in the resolution of signal spectra in the time–frequency plane.

The Wigner–Ville distribution (WVD), after Eugene Paul Wigner (1932) and Jean-André Ville (1948), is one of such quadratic representations for nonstationary signal analysis in the time–frequency plane.

6.1 Basics of the Wigner–Ville Distribution (WVD)

A seismic trace is fundamentally a form of wavelet collection that follows the principle of superposition. The WVD method can facilitate the representation of the distribution of the individual wavelet energy packets.

The WVD has two basic features: quadratic representation and quasi-probability distribution. The WVD is a quadratic representation of the

Time–Frequency Analysis of Seismic Signals, First Edition. Yanghua Wang.

instantaneous time–frequency energy of a signal. The energy E_x of a signal can be deduced from the squared modulus of either the signal itself $x(t)$ or its Fourier transform $\hat{x}(f)$:

$$E_x = \int_{-\infty}^{\infty} |x(t)|^2 \mathrm{d}t = \int_{-\infty}^{\infty} |\hat{x}(f)|^2 \mathrm{d}f, \tag{6.1}$$

where $|x(t)|^2$ can be interpreted as the instantaneous energy in the time domain, and $|\hat{x}(f)|^2$ is the spectral energy density in the frequency domain. It is natural to look for a joint time–frequency energy density $p_x(t,f)$, which combines the concepts of the instantaneous energy and the spectral energy density (Ville, 1948).

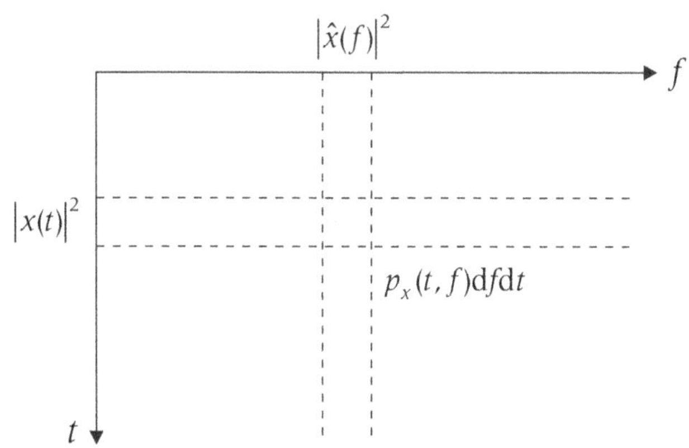

Figure 6.1 The joint time–frequency energy density $p_x(t,f)$. The horizontal band is the distribution of spectral energy density along the frequency axis, and when integrated, the result is $|x(t)|^2$. The vertical band is the distribution of instantaneous energy along the time axis, and when integrated, the result is $|\hat{x}(f)|^2$.

Figure 6.1 illustrates the energy density distribution $p_x(t,f)$ in the time–frequency plane. The horizontal band, at any time t, gives the distribution of the spectral energy density over frequency f. The vertical band is the distribution of the instantaneous energy, carried by one frequency f.

The energy of a two-dimensional element $\mathrm{d}f\mathrm{d}t$ is $p_x(t,f)\mathrm{d}f\mathrm{d}t$. Therefore, the total energy E_x of signal $x(t)$ is

$$E_x = \int_{-\infty}^{\infty}\int_{-\infty}^{\infty} p_x(t,f)\mathrm{d}f\mathrm{d}t. \tag{6.2}$$

Since energy is a quadratic function of the signal, the energy density distribution could generally be a quadratic representation in the time–frequency plane. The quadratic representation is one of the motivations for the development of WVD mentioned above (Page, 1952; Cohen, 1989; Hlawatsch & Boudreaux-Bartels, 1992; Debnath & Shah, 2015)

The second basic feature of WVD is the quasi-probability distribution, which mathematically resembles a probability distribution but relaxes some of the Kolmogorov axioms of probability theory (Kolmogorov, 1950). For example, as a quasi-probability distribution, the WVD spectrum counter-intuitively has regions of negative energy density $p_x(t,f)$, which contradicts the Kolmogorov axiom that the probability of an event is a non-negative real number.

For the WVD, there are three important properties:

(1) Marginality.
(2) Unitarity.
(3) Frequency-support conservation.

The marginal property is that, as illustrated in Figure 6.1, the joint energy density $p_x(t,f)$ satisfies the time and frequency marginal properties:

$$\int_{-\infty}^{\infty} p_x(t,f)\mathrm{d}f = |x(t)|^2, \tag{6.3}$$

$$\int_{-\infty}^{\infty} p_x(t,f)\mathrm{d}t = |\hat{x}(f)|^2. \tag{6.4}$$

If the time–frequency energy density is integrated along one variable, the result is the energy density corresponding to the other variable.

The unitarity expresses energy conservation from the (squared) time-domain scalar product to the product in the time–frequency domain:

$$\left|\int_{-\infty}^{\infty} x(t)y^*(t)\mathrm{d}t\right|^2 = \int_{-\infty}^{\infty}\int_{-\infty}^{\infty} p_x(t,f)p_y^*(t,f)\mathrm{d}f\mathrm{d}t. \tag{6.5}$$

The support conservation can be analysed on the time and frequency axes separately. If a signal has a compact support in the time domain, its time–frequency spectrum also has the same compact support on the time axis. An attractive property of the WVD method is its frequency-support conservation. If a signal has a compact support in the frequency domain, its spectrum also has the same compact support on the frequency axis:

$$\begin{aligned} x(t) = 0, \quad |t| > T, \quad &\Rightarrow \quad p_x(t,f) = 0, \quad |t| > T, \\ \hat{x}(f) = 0, \quad |f| > B, \quad &\Rightarrow \quad p_x(t,f) = 0, \quad |f| > B. \end{aligned} \tag{6.6}$$

In contrast, the Gabor transform type methods do not even have such a weak finite support, although all these time–frequency spectral analysis methods follow the Gabor limit (Eq. 2.28) for the maximum possible resolution.

6.2 Defining the WVD with an Analytic Signal

In addition to the energetic interpretation, another motivation for the WVD development is the correlative interpretation.

The motivation of correlative interpretation leads to the definition of the WVD, as (Wigner, 1932)

$$p_x(t,f) = \int_{-\infty}^{\infty} x\left(t + \frac{\tau}{2}\right) x^*\left(t - \frac{\tau}{2}\right) e^{-i2\pi f\tau} d\tau, \tag{6.7}$$

where the WVD kernel $x(t+\tau/2)x^*(t-\tau/2)$ is the instantaneous auto-correlation, and τ is the correlation time lag (Mark, 1970). It can be proved that the WVD definition in Eq. (6.7) satisfies the marginal properties of Eqs. (6.3–6.4). The validity is summarised in Appendix E.1.

However, the WVD of a real signal $x(t)$ does not, in general, represent the true time–frequency pattern of the signal. Let us demonstrate this point using a linear chirp signal:

$$x(t) = b(t)\cos\left[2\pi\left(f_0 t + \frac{\beta}{2}t^2\right)\right], \tag{6.8}$$

where $b(t)$ is the boxcar window function:

$$b(t) = \begin{cases} 1, & \text{for } |t| \le T, \\ 0, & \text{otherwise,} \end{cases} \tag{6.9}$$

and β is the frequency change rate along the time axis, which defines the instantaneous frequency of the signal as (Eq. 1.34)

$$f_{\text{inst}}(t) = \frac{1}{2\pi}\frac{\mathrm{d}\theta(t)}{\mathrm{d}t} = f_0 + \beta t, \tag{6.10}$$

for $|t| \leq T$. The frequency varies within the range $f \leq f_{\text{inst}}(t) \pm B$, where B is a constant (half-)bandwidth. Figure 6.2a shows an example of the linear chirp signal $x(t)$, that is defined with parameters $f_0 = 30\,\text{Hz}$, $\beta = 100\,\text{Hz/s}$, and $T = 0.3\,\text{s}$.

The WVD of this linear chirp signal $x(t)$ is (Appendix E.2)

$$p_x(t,f) \approx (T - |t|)\left(\frac{\sin[4\pi[f - f_{\text{inst}}(t)](T - |t|)]}{4\pi[f - f_{\text{inst}}(t)](T - |t|)} + \frac{\sin[4\pi[f + f_{\text{inst}}(t)](T - |t|)]}{4\pi[f + f_{\text{inst}}(t)](T - |t|)}\right)$$
$$+\frac{1}{\sqrt{2\beta}}\cos\left[2\pi\left([f_0 + f_{\text{inst}}(t)]t - \frac{f^2}{\beta}\right) + \frac{\pi}{4}\right], \tag{6.11}$$

for $|t| \leq T$, and $|f| \leq f_{\text{inst}}(t) + B$. The two sinc functions, $\sin(u)/u$, indicate that there are two peak values, at $f = \pm f_{\text{inst}}(t)$, respectively, as illustrated in Figure 6.2b. However, the energy concentration along $f = -f_{\text{inst}}(t)$ should not be exhibited, and this energy concentration is an artefact that is generated by the WVD.

It is advisable to produce an energy density distribution without any artefact appearing on the negative frequency side. For this purpose, the analytic signal $z(t)$, rather than the real signal $x(t)$ itself, can be used for the instantaneous autocorrelation, since the Fourier transform $\hat{z}(f)$ of the analytic signal $z(t)$ has no negative frequency components (Eq. 1.24). Consequently, there is no interference between the positive and negative frequency components. The WVD, defined in terms of the analytic signal $z(t)$, is

$$S_{\text{WVD}}(t,f) = \int_{-\infty}^{\infty} z\left(t + \frac{\tau}{2}\right) z^*\left(t - \frac{\tau}{2}\right) \mathrm{e}^{-\mathrm{i}2\pi f\tau}\,\mathrm{d}\tau. \tag{6.12}$$

This WVD definition was originally proposed by Ville (1948).

Applying Parseval's theorem, the WVD formula can also be written in terms of the frequency-domain analytic signal $\hat{z}(f)$ as

$$S_{\text{WVD}}(t,f) = \int_{-\infty}^{\infty} \hat{z}\left(f + \frac{\xi}{2}\right) \hat{z}^*\left(f - \frac{\xi}{2}\right) \mathrm{e}^{\mathrm{i}2\pi\xi t}\,\mathrm{d}\xi. \tag{6.13}$$

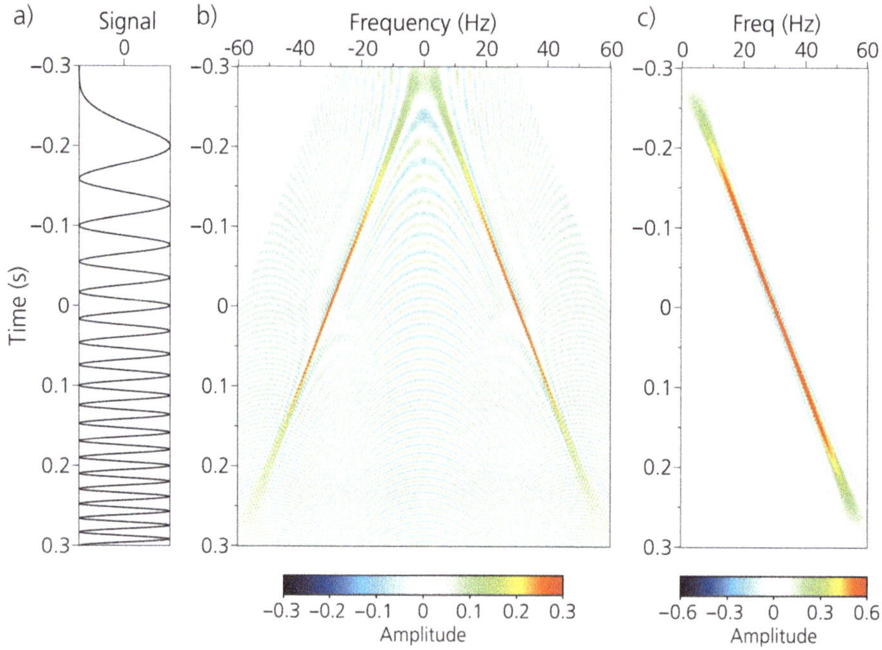

Figure 6.2 WVDs defined by the real and the analytic signals. (a) A linear chirp signal $x(t)$. (b) The WVD $p_x(t, f)$ of the real signal $x(t)$. (c) The WVD $S_{\mathrm{WVD}}(t, f)$ of the analytic signal $z(t)$.

For the sample linear chirp signal of Eq. (6.8), the WVD of the analytic signal $z(t)$ can be expressed as (Appendix E.2)

$$S_{\mathrm{WVD}}(t,f) = 4(T-|t|)\left(\frac{\sin\left[4\pi[f-f_{\mathrm{inst}}(t)](T-|t|)\right]}{4\pi[f-f_{\mathrm{inst}}(t)](T-|t|)}\right), \tag{6.14}$$

for $|t| \le T$, and $f \le f_{\mathrm{inst}}(t) \pm B$. This WVD, as shown in Figure 6.2c, clearly exhibits energy concentration only at the instantaneous frequency $f_{\mathrm{inst}}(t)$ of the original signal $x(t)$, because the analytic signal $z(t)$ annuls negative frequencies of the Fourier transform.

Using Eq. (6.14), the WVD of the real signal $x(t)$ (Eq. 6.11) can be rewritten as

$$p_x(t,f) \approx \frac{1}{4}\left[S_{\mathrm{WVD}}(t,f) + S_{\mathrm{WVD}}(t,-f)\right] + \frac{1}{\sqrt{2\beta}}\cos\left[2\pi\left([f_0 + f_{\mathrm{inst}}(t)]t - \frac{f^2}{\beta}\right) + \frac{\pi}{4}\right], \tag{6.15}$$

for $|t| \le T$, and $|f| \le f_{\text{inst}}(t) + B$. The WVD $p_x(t, f)$ in Eq. (6.15) has three terms. The first two terms, $S_{\text{WVD}}(t, f)$ and $S_{\text{WVD}}(t, -f)$, are produced by two conjugated analytic signals, because any real signal $x(t)$ is equivalent to a sum of an analytic signal and its conjugate. However, the WVD $p_x(t, f)$ also shows a third term, which is the interference between those two conjugated analytic signals.

In summary, the WVD should be calculated based on the analytic signal $z(t)$, rather than the original real signal $x(t)$.

6.3 Properties of the WVD

Although the WVD of the analytic signal $z(t)$ is defined in terms of a complex autocorrelation, the WVD spectrum is in real:

$$\begin{aligned} S^*_{\text{WVD}}(t, f) &= \int_{-\infty}^{\infty} z^*\left(t + \frac{\tau}{2}\right) z\left(t - \frac{\tau}{2}\right) e^{i2\pi f\tau} d\tau \\ &= \int_{-\infty}^{\infty} z^*\left(t - \frac{u}{2}\right) z\left(t + \frac{u}{2}\right) e^{-i2\pi fu} du = S_{\text{WVD}}(t, f), \end{aligned} \tag{6.16}$$

where $u = -\tau$. Therefore, the WVD does not have the phase information. Figure 6.2c demonstrates that the WVD function is in real, although it is not necessarily non-negative. Both the lack of phase information and the fact of containing negative values are two drawbacks of the WVD method.

The WVD of the analytic signal $z(t)$ satisfies the time and frequency marginal properties:

$$\int_{-\infty}^{\infty} S_{\text{WVD}}(t, f) df = |z(t)|^2, \tag{6.17}$$

$$\int_{-\infty}^{\infty} S_{\text{WVD}}(t, f) dt = |\hat{z}(f)|^2. \tag{6.18}$$

They are in parallel to Eqs. (6.3) and (6.4), which are for the WVD of the real signal $x(t)$. The time marginal property can be proved, based on the WVD definition in Eq. (6.12):

$$\int_{-\infty}^{\infty} S_{\text{WVD}}(t, f) e^{i2\pi f\tau} df = z\left(t + \frac{\tau}{2}\right) z^*\left(t - \frac{\tau}{2}\right). \tag{6.19}$$

Setting $t_1 = t + \tau/2$ and $t_2 = t - \tau/2$, Eq. (6.19) becomes

$$\int_{-\infty}^{\infty} S_{\mathrm{WVD}}\left(\frac{t_1 + t_2}{2}, f\right) \mathrm{e}^{\mathrm{i}2\pi f(t_1 - t_2)} \mathrm{d}f = z(t_1) z^*(t_2). \tag{6.20}$$

If $t_1 = t_2 = t$, this equation leads to the time marginal property in Eq. (6.17). That is, the frequency integral of the WVD corresponds to the signal's instantaneous energy in time.

The frequency marginal property can also be proved by, using the WVD definition in Eq. (6.13):

$$\int_{-\infty}^{\infty} S_{\mathrm{WVD}}(t, f) \mathrm{e}^{-\mathrm{i}2\pi\xi t} \mathrm{d}t = \hat{z}\left(f + \frac{\xi}{2}\right) \hat{z}^*\left(f - \frac{\xi}{2}\right), \tag{6.21}$$

$$\int_{-\infty}^{\infty} S_{\mathrm{WVD}}\left(t, \frac{f_1 + f_2}{2}\right) \mathrm{e}^{-\mathrm{i}2\pi(f_1 - f_2)t} \mathrm{d}t = \hat{z}(f_1) \hat{z}^*(f_2). \tag{6.22}$$

Setting $f_1 = f_2 = f$ gives the frequency marginal property that is presented in Eq. (6.18). That is, the time integral of the WVD corresponds to the spectral energy density.

Since the WVD satisfies the marginal properties, it also fulfils the total energy requirement (Cohen, 1989):

$$E_z = \int_{-\infty}^{\infty} \int_{-\infty}^{\infty} S_{\mathrm{WVD}}(t, f) \mathrm{d}f \mathrm{d}t = \int_{-\infty}^{\infty} |z(t)|^2 \mathrm{d}t = \int_{-\infty}^{\infty} |\hat{z}(f)|^2 \mathrm{d}f. \tag{6.23}$$

The instantaneous frequency of an analytic signal $z(t)$ can be recovered from the WVD as its first order moment (or gravitational centre) in the frequency domain (Claasen & Meckidnbräuker, 1980a, 1981):

$$f_c(t) = \frac{\int_{-\infty}^{\infty} f S_{\mathrm{WVD}}(t, f) \mathrm{d}f}{\int_{-\infty}^{\infty} S_{\mathrm{WVD}}(t, f) \mathrm{d}f} = \frac{1}{2\pi} \mathrm{Im}\left\{\frac{z'(t)}{z(t)}\right\}, \tag{6.24}$$

where $z'(t)$ is the time derivative of $z(t)$. If $z(t) = a(t)\mathrm{e}^{\mathrm{i}\theta(t)}$, Eq. (6.24) becomes $f_c(t) = (2\pi)^{-1} \mathrm{d}\theta(t)/\mathrm{d}t$, the instantaneous frequency defined in Eq. (1.34).

In the same way, the group delay of the signal can be recovered from the WVD as its first order moment in time:

$$t_c(f) = \frac{\int_{-\infty}^{\infty} t S_{\mathrm{WVD}}(t,f)\mathrm{d}t}{\int_{-\infty}^{\infty} S_{\mathrm{WVD}}(t,f)\mathrm{d}t} = -\frac{1}{2\pi}\mathrm{Im}\left\{\frac{\hat{z}'(f)}{\hat{z}(f)}\right\}. \quad (6.25)$$

Linear operations on the analytic signal $z(t)$, such as filtering and multiplication, are directly mirrored in linear operations on the WVD of the signal (Claasen & Meckidnbräuker, 1980a). Explicitly, if an analytic signal $y(t)$ is convolved with a filter $h(t)$, then the WVD representation maintains time linearity, thus:

$$z(t) = \int_{-\infty}^{\infty} h(t-\tau)y(\tau)\mathrm{d}\tau \quad \Leftrightarrow \quad S_{\mathrm{WVD}}(t,f) = \int_{-\infty}^{\infty} S_{\mathrm{WVD},h}(t-\tau,f)S_{\mathrm{WVD},y}(\tau,f)\mathrm{d}\tau, \quad (6.26)$$

where $S_{\mathrm{WVD},y}(t,f)$ and $S_{\mathrm{WVD},h}(t,f)$ are the WVDs of $y(t)$ and $h(t)$, respectively. Similarly, a time-domain multiplication becomes a frequency-domain convolution:

$$z(t) = h(t)y(t) \quad \Leftrightarrow \quad S_{\mathrm{WVD}}(t,f) = \int_{-\infty}^{\infty} S_{\mathrm{WVD},h}(t,f-\xi)S_{\mathrm{WVD},y}(t,\xi)\mathrm{d}\xi. \quad (6.27)$$

It should be noted that for most other time–frequency representations, including the STFT, such relations do not exist.

However, the WVD represents an energy density distribution, from which it is not possible to recover the signal uniquely. Based on the WVD definition, the instantaneous autocorrelation can be recovered by an inverse Fourier transform:

$$z\left(t+\frac{\tau}{2}\right)z^*\left(t-\frac{\tau}{2}\right) = \int_{-\infty}^{\infty} S_{\mathrm{WVD}}(t,f)\mathrm{e}^{\mathrm{i}2\pi f\tau}\mathrm{d}f. \quad (6.28)$$

At the specific value $t = \tau/2$, it can be expressed as

$$z(\tau) = \frac{1}{z^*(0)}\int_{-\infty}^{\infty} S_{\mathrm{WVD}}\left(\frac{\tau}{2},f\right)\mathrm{e}^{\mathrm{i}2\pi f\tau}\mathrm{d}f. \quad (6.29)$$

Hence, the analytic signal $z(t)$ can only be recovered up to a constant factor $|z(0)|$. The phase corresponding to the constant factor, however, can never be recovered.

6.4 The Smoothed WVD

For an analytic signal $z(t)$ composited of M signal components,

$$z(t) = \sum_{m=1}^{M} c_m z_m(t), \tag{6.30}$$

where c_m are the real coefficients, the instantaneous autocorrelation consists of not only the autocorrelation terms of each individual signal component but also the cross-correlation terms between any pair of signal components. Therefore, the WVD is

$$S_{\mathrm{WVD}}(t,f) = \sum_{m=1}^{M} c_m^2 P_{z_m}(t,f) + \sum_{m=1}^{M-1}\sum_{n=m+1}^{M} c_m c_n I_{z_m,z_n}(t,f), \tag{6.31}$$

where $P_{z_m}(t,f)$ is the auto-term WVD, which is defined by the instantaneous autocorrelation of the signal component, and $I_{z_m,z_n}(t,f)$ is the cross-term interference between any two components (Claasen & Meckidnbräuker, 1980a):

$$I_{z_m,z_n}(t,f) = Q_{z_m,z_n}(t,f) + Q_{z_n,z_m}(t,f). \tag{6.32}$$

Each cross-term interference is formed by a pair of cross-term WVDs, defined in terms of the instantaneous cross-correlation, as

$$Q_{z_m,z_n}(t,f) = \int_{-\infty}^{\infty} z_m\left(t+\frac{\tau}{2}\right) z_n^*\left(t-\frac{\tau}{2}\right) \mathrm{e}^{-\mathrm{i}2\pi f\tau}\mathrm{d}\tau. \tag{6.33}$$

The location of each cross-term interference $I_{z_m,z_n}(t,f)$ in the time–frequency plane is at the geometrical midpoint of the two signal components $[z_m(t), z_n(t)]$. Let us demonstrate this geometrical feature with the following two signal components:

$$\begin{aligned} z_m &= g(t-t_m)\mathrm{e}^{\mathrm{i}(2\pi f_m t+\theta_m)}, \\ z_n &= g(t-t_n)\mathrm{e}^{\mathrm{i}(2\pi f_n t+\theta_n)}, \end{aligned} \tag{6.34}$$

where $g(t)$ is a time window function. The two auto-term WVDs are

$$\begin{aligned} P_{z_m}(t,f) &= P_g(t-t_m, f-f_m), \\ P_{z_n}(t,f) &= P_g(t-t_n, f-f_n). \end{aligned} \tag{6.35}$$

The two cross-term WVDs are

$$Q_{z_m,z_n}(t,f)=\int_{-\infty}^{\infty} g\left(t-t_0+\frac{\tau-\Delta t}{2}\right) g\left(t-t_0-\frac{\tau-\Delta t}{2}\right) \times \mathrm{e}^{\mathrm{i}(2\pi t\Delta f+\Delta\theta)}\mathrm{e}^{-\mathrm{i}2\pi(f-f_0)\tau}\mathrm{d}\tau = P_g(t-t_0,f-f_0)\mathrm{e}^{\mathrm{i}2\pi[(t-t_0)\Delta f-(f-f_0)\Delta t]}\mathrm{e}^{\mathrm{i}(2\pi t_0\Delta f+\Delta\theta)}, \tag{6.36}$$

$$Q_{z_n,z_m}(t,f)=P_g(t-t_0,f-f_0)\mathrm{e}^{-\mathrm{i}2\pi[(t-t_0)\Delta f-(f-f_0)\Delta t]}\mathrm{e}^{-\mathrm{i}(2\pi t_0\Delta f+\Delta\theta)}, \tag{6.37}$$

where $t_0=(t_m+t_n)/2$, $\Delta t=t_m-t_n$, $f_0=(f_m+f_n)/2$, $\Delta f=f_m-f_n$, and $\Delta\theta=\theta_m-\theta_n$. Therefore, the cross-term interference is

$$I_{z_m,z_n}(t,f)=2P_g(t-t_0,f-f_0) \times\cos\left[2\pi(t-t_0)\Delta f-2\pi(f-f_0)\Delta t+(2\pi t_0\Delta f+\Delta\theta)\right]. \tag{6.38}$$

Suppose the time window is in a Gaussian form $g(t)=\mathrm{e}^{-t^2/(2\sigma^2)}$, the auto-term WVD is

$$P_g(t,f)=2\sqrt{\pi}\sigma\mathrm{e}^{-\frac{t^2}{\sigma^2}-4\pi^2\sigma^2f^2}. \tag{6.39}$$

Figure 6.3 displays the WVD of a signal composed of two Gabor wavelets:

$$z(t)=c_mz_m(t)+c_nz_n(t) = c_m\mathrm{e}^{-\frac{(t-t_m)^2}{2\sigma^2}}\mathrm{e}^{\mathrm{i}(2\pi f_mt+\theta_m)}+c_n\mathrm{e}^{-\frac{(t-t_n)^2}{2\sigma^2}}\mathrm{e}^{\mathrm{i}(2\pi f_nt+\theta_n)}, \tag{6.40}$$

where $\sigma=0.05\mathrm{s}$, $(c_m,c_n)=(1.0,0.7)$, $(t_m,t_n)=(0.15\mathrm{s},0.45\mathrm{s})$, $(f_m,f_n)=(50\,\mathrm{Hz}, 20\,\mathrm{Hz})$, and $\theta_m=\theta_n=0$. The WVD clearly shows that

(1) The cross-term interference is at the midpoint, and at $(t_0,f_0)=(0.3\mathrm{s}, 35\,\mathrm{Hz})$ in this case.
(2) The cross-term interference oscillates perpendicularly to the line joining the two points.
(3) The cross-term interference is oscillatory with positive and negative values.

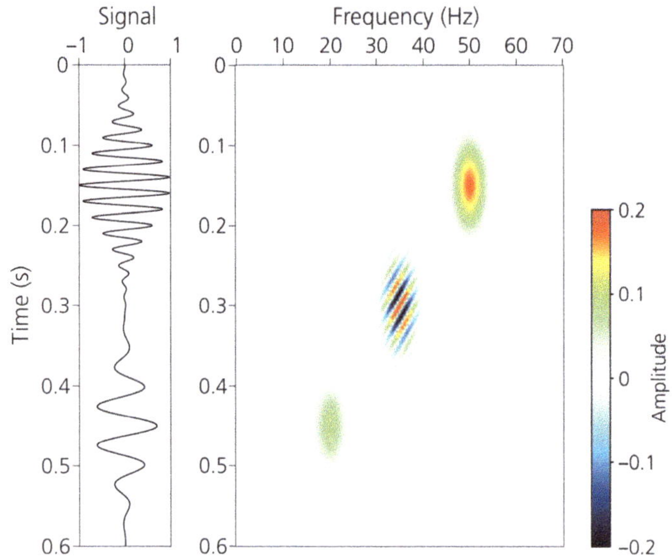

Figure 6.3 The cross-term interference in the WVD. The location of the cross-term interference in the time–frequency plane is at the geometric midpoint of each pair of components.

Since the cross-term interference is oscillatory with positive and negative values in the time–frequency plane, it can be easily suppressed by smoothing. The smoothing process includes time-axis smoothing and frequency-axis smoothing. The time-smoothing filter $g(t)$ and the frequency-smoothing filter $\hat{h}(f)$ can be defined in Gaussian forms as (Shin & Jeon, 1993),

$$g(t) = \frac{1}{\sqrt{2\pi}\sigma_t} e^{-\frac{t^2}{2\sigma_t^2}}, \tag{6.41}$$

$$\hat{h}(f) = \frac{1}{\sqrt{2\pi}\sigma_f} e^{-\frac{f^2}{2\sigma_f^2}}, \tag{6.42}$$

where σ_t and σ_f are the spread parameters controlling the breadth of the Gaussian functions in the time and frequency axes, respectively. These two one-dimensional filters formulate a two-dimensional smoothing filter $K(t, f)$ as

$$K(t, f) = g(t)\hat{h}(f) = \frac{1}{2\pi\sigma_f\sigma_t} e^{-\frac{f^2}{2\sigma_f^2}-\frac{t^2}{2\sigma_t^2}}. \tag{6.43}$$

Therefore, the smoothing process can be expressed as a two-dimensional convolution in the time–frequency plane:

$$\tilde{S}_{\mathrm{WVD}}(t,f) = \int_{-\infty}^{\infty}\int_{-\infty}^{\infty} K(t-t', f-f')S_{\mathrm{WVD}}(t', f')\mathrm{d}f'\mathrm{d}t'. \tag{6.44}$$

Now, let us discuss the frequency smoothing in the time domain, rather than in the frequency domain directly. The frequency-domain smoothing process

$$\tilde{S}_{\mathrm{WVD}}(t,f) = \int_{-\infty}^{\infty} \hat{h}(f-\xi)S_{\mathrm{WVD}}(t,\xi)\mathrm{d}\xi \tag{6.45}$$

is equivalent to a time-domain windowing, which can be expressed as

$$\begin{aligned}\tilde{S}_{\mathrm{WVD}}(t,f) &= \int_{-\infty}^{\infty} h(\tau) z\left(t+\frac{\tau}{2}\right) z^*\left(t-\frac{\tau}{2}\right) \mathrm{e}^{-\mathrm{i}2\pi f\tau}\mathrm{d}\tau \\ &= \int_{-\infty}^{\infty} k\left(\frac{\tau}{2}\right) k^*\left(-\frac{\tau}{2}\right) z\left(t+\frac{\tau}{2}\right) z^*\left(t-\frac{\tau}{2}\right) \mathrm{e}^{-\mathrm{i}2\pi f\tau}\mathrm{d}\tau. \end{aligned} \tag{6.46}$$

Here, the frequency-smoothing filter $h(\tau) = k(\tau/2)k^*(-\tau/2)$ is formed as the square of the time window function $k(\tau/2)$. Considering $k(\tau/2) = k(t=\tau, \tau/2)$, Eq. (6.46) is the WVD of a windowed signal $k(t)z(t)$. The WVD of this time-windowed signal is called the pseudo-WVD, which is distinguished from the classic WVD that is defined by the signal with an infinite time duration (Claasen & Mecklenbräuker, 1980a).

The frequency-smoothing filter $\hat{h}(f)$ of Eq. (6.42) can be expressed in the time domain, as

$$h(t) = \mathrm{e}^{-2\pi^2\sigma_f^2 t^2}, \tag{6.47}$$

and thus, the time window function $k(t)$ is

$$k(t) = \mathrm{e}^{-4\pi^2\sigma_f^2 t^2}. \tag{6.48}$$

It is easy to prove that the frequency-smoothing filter $\hat{h}(f)$ is the WVD of the time window function $k(t)$, as

$$\hat{h}(f) = \int_{-\infty}^{\infty} k\left(\frac{\tau}{2}\right) k^*\left(-\frac{\tau}{2}\right) e^{-i2\pi f\tau} d\tau. \tag{6.49}$$

Then, the time-smoothing filter is also applied, to smooth the pseudo-WVD of Eq. (6.46) along the time axis. The final smoothed WVD becomes

$$\tilde{S}_{\mathrm{WVD}}(t,f) = \int_{-\infty}^{\infty} g(t-t') \int_{-\infty}^{\infty} k\left(\frac{\tau}{2}\right) z\left(t'+\frac{\tau}{2}\right) k^*\left(-\frac{\tau}{2}\right) z^*\left(t'-\frac{\tau}{2}\right) e^{-i2\pi f\tau} d\tau dt'. \tag{6.50}$$

This time–frequency smoothed WVD is often called 'the smoothed pseudo-WVD'. Here, we simply call it 'the smoothed WVD' because, a seismic trace practically always has a finite time duration, and the WVD is always a pseudo-WVD (Flandrin & Escudié, 1984).

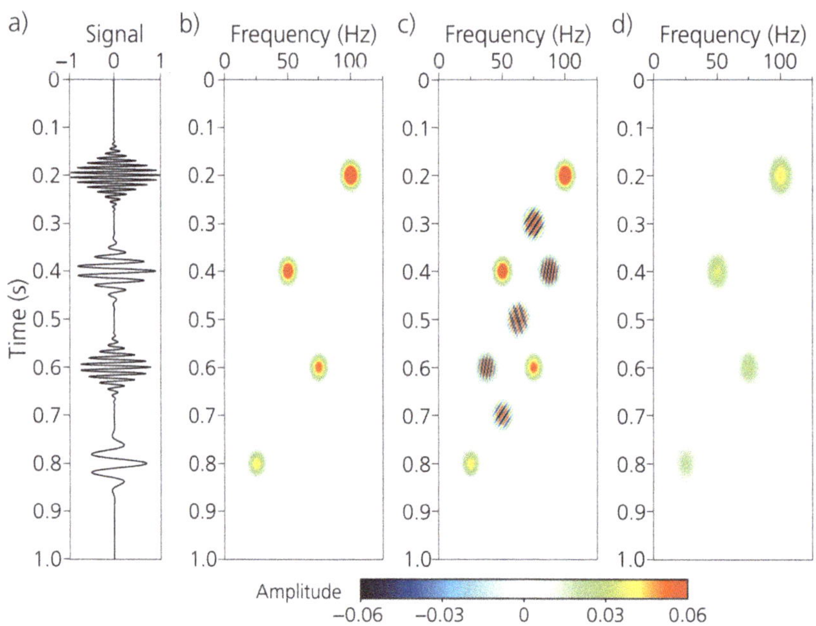

Figure 6.4 The smoothing effect in the WVD. (a) A synthetic seismic trace consisting of four Gabor wavelets. (b) The WVD representation of the auto-terms of four Gabor wavelets. (c) The WVD contains not only the auto-term representations but also the cross-term interference between any pair of wavelets. (d) The smoothed WVD suppresses the interference but also reduces the resolution.

Figure 6.4a shows another WVD example, in which the number of Gabor wavelets in the synthetic seismic trace is increased from two in Figure 6.3 to four ($M = 4$). The central frequencies of these four wavelets are $f_m = \{100, 50, 75, 25\}$ Hz, and the corresponding amplitudes are $c_m = \{1.0, 0.9, 0.8, 0.7\}$. The Gaussian spread parameter σ is a constant, $\sigma = 0.015$ s. The WVD, according to Eq. (6.31), is

$$S_{\text{WVD}}(t,f) = 2\sqrt{\pi}\sigma \sum_{m=1}^{M} c_m^2 \, e^{-\frac{t_m^2}{\sigma^2} - 4\pi^2\sigma^2 f_m^2}$$

$$+2\sum_{m=1}^{M-1}\sum_{n=m+1}^{M} c_m c_n \, e^{-\frac{1}{\sigma^2}\left(t-\frac{t_m+t_n}{2}\right)^2 - 4\pi^2\sigma^2\left(f-\frac{f_m+f_n}{2}\right)^2}$$

$$\times\cos\left[2\pi\left(t-\frac{t_m+t_n}{2}\right)(f_m-f_n) - 2\pi\left(f-\frac{f_m+f_n}{2}\right)(t_m-t_n)\right.$$

$$\left. +2\pi\left(\frac{t_m+t_n}{2}\right)(f_m-f_n) + (\theta_m-\theta_n)\right]. \tag{6.51}$$

Figure 6.4b displays the auto-terms of the WVD corresponding to the four Gabor wavelets. But as shown in Figure 6.4c, the WVD given by Eq. (6.51) also contains the cross-term interference.

Figure 6.4d demonstrates the effectiveness of the two-dimensional smoothing process. The two parameters that determine the time and frequency resolution of the smoothing filter $K(t,f)$ are set as $\sigma_t = 0.02$ s and $\sigma_f = 4$ Hz, respectively. While the smoothed WVD suppresses the cross-term interference, the convolution operation for smoothing obviously also compromises the energy concentration of the auto-terms in the WVD and reduces the time–frequency resolution.

Although one can visually judge whether the cross-term interference has been suppressed after the smoothing process, we can select the smoothing parameters (σ_t, σ_f) for the two-dimensional filter $K(t,f)$ quantitatively, by measuring the negative values in the smoothed WVD as

$$r^{(-)}(\sigma_t,\sigma_f) = 20\log_{10}\frac{\sum \tilde{S}_{\text{WVD}}^{(-)}(t,f;\sigma_t,\sigma_f)}{\sum S_{\text{WVD}}^{(-)}(t,f)}, \tag{6.52}$$

where $S_{\text{WVD}}^{(-)}(t,f)$ and $\tilde{S}_{\text{WVD}}^{(-)}(t,f;\sigma_t,\sigma_f)$ are negative values in the original and smoothed WVD, and $r^{(-)}(\sigma_t,\sigma_f)$ with the superscript '(−)' emphasises that the measurement (in dB) is made from the negative WVD values.

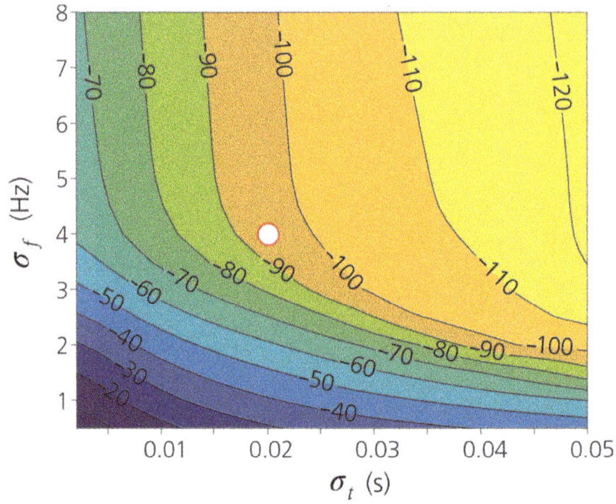

Figure 6.5 Selecting the parameter pair (σ_t, σ_f) for the smoothing filter. The colour scale represents the magnitude of the sum of the negative values in the smoothed WVD, and the magnitude is normalised to dB.

Figure 6.5 displays the measurement $r^{(-)}(\sigma_t, \sigma_f)$, for the quantitative selection of the parameter pair (σ_t, σ_f), for the example shown in Figure 6.4. The selected (white) point $(\sigma_t, \sigma_f) = (0.02\,\text{s}, 4\,\text{Hz})$ is at the contour of -95 dB, and it has the shortest distance to the origin.

6.5 The Generalised Class of Time–Frequency Representations

The time–frequency smoothing process in Eq. (6.44) is a two-dimensional convolution. This convolutional expression is a generalised class of time–frequency representations, $C_z(t, f)$, that was introduced by Cohen (1966, 1989):

$$C_z(t, f) = \int_{-\infty}^{\infty} \int_{-\infty}^{\infty} K(t - t', f - f') S_{\text{WVD}}(t', f') \text{d}f' \text{d}t', \tag{6.53}$$

where $K(t, f)$ is a generalised kernel that is applied to the time–frequency spectrum $S_{\text{WVD}}(t, f)$.

According to the Fourier theorem, if $z(t)$ is scaled by a, the Fourier transform $\hat{z}(f) = \mathcal{F}[z(t)]$ is scaled by a^{-1}. The same should be reflected in the joint energy distribution $C_z(t, f)$ as well:

$$z_a(t) = \frac{1}{\sqrt{a}} z\left(\frac{t}{a}\right) \quad \Leftrightarrow \quad C_{z_a}(t, f) = C_z\left(\frac{t}{a}, af\right). \tag{6.54}$$

The generalised kernel $K(t, f)$ ensures that the generalised class of time–frequency representations satisfy the time translation and frequency modulation invariance properties:

$$z_b(t) = z(t - t_0)\mathrm{e}^{\mathrm{i}2\pi f_0 t} \quad \Leftrightarrow \quad C_{z_b}(t, f) = C_z(t - t_0, f - f_0). \tag{6.55}$$

The generalised class of time–frequency representations $C_z(t, f)$ can also be understood as the Fourier transform of the following generalised autocorrelation function:

$$R(t, \tau) = \int_{-\infty}^{\infty} \gamma(t - t', \tau) z\left(t' + \frac{\tau}{2}\right) z^*\left(t' - \frac{\tau}{2}\right) \mathrm{d}t', \tag{6.56}$$

where $\gamma(t, \tau)$ is the parameterisation kernel at time τ. Once the parameterisation kernel is applied as a convolution operator to the regular autocorrelation, $R(t, \tau)$ at a given time τ becomes the generalised form of the autocorrelation function, rather than the ordinary autocorrelation that is used in the WVD.

The Fourier transform of $R(t, \tau)$ with respect to time τ is

$$C_z(t, f) = \int_{-\infty}^{\infty} \int_{-\infty}^{\infty} \gamma(t - t', \tau) z\left(t' + \frac{\tau}{2}\right) z^*\left(t' - \frac{\tau}{2}\right) \mathrm{e}^{-\mathrm{i}2\pi f \tau} \mathrm{d}t' \mathrm{d}\tau. \tag{6.57}$$

Therefore, the generalised kernel $K(t, f)$ can be built from the parameterisation kernel $\gamma(t, \tau)$ by the Fourier transform:

$$K(t, f) = \int_{-\infty}^{\infty} \gamma(t, \tau) \mathrm{e}^{-\mathrm{i}2\pi f \tau} \mathrm{d}\tau = \int_{-\infty}^{\infty} \int_{-\infty}^{\infty} \hat{\gamma}(\xi, \tau) \mathrm{e}^{\mathrm{i}2\pi(\xi t - f\tau)} \mathrm{d}\tau \mathrm{d}\xi, \tag{6.58}$$

where $\hat{\gamma}(\xi, \tau)$ is the parameterisation kernel in the frequency domain, at a given time τ. Given the parameterisation kernel $\hat{\gamma}(\xi, \tau)$ as a representative of a particular distribution function, the generalised class of time–frequency representations of Eq. (6.57) can be written as

$$C_z(t, f) = \int_{-\infty}^{\infty} \int_{-\infty}^{\infty} \int_{-\infty}^{\infty} \hat{\gamma}(\xi, \tau) z\left(t' + \frac{\tau}{2}\right) z^*\left(t' - \frac{\tau}{2}\right) \mathrm{e}^{\mathrm{i}2\pi[\xi(t - t') - f\tau]} \mathrm{d}t' \mathrm{d}\tau \mathrm{d}\xi. \tag{6.59}$$

If the parameterisation kernel $\hat{\gamma}(\xi, \tau)$ is set as an exponential function (Choi & Williams, 1989):

$$\hat{\gamma}(\xi, \tau) = \mathrm{e}^{-\frac{\xi^2 \tau^2}{2\sigma^2}}, \tag{6.60}$$

then, the parameterisation kernel $\gamma(t, \tau)$ in the time domain is

$$\gamma(t, \tau) = \mathcal{F}_{\xi}^{-1}\left[\hat{\gamma}(\xi, \tau)\right] = \frac{\sqrt{2\pi}\sigma}{\tau} \mathrm{e}^{-\frac{2\pi^2\sigma^2 t^2}{\tau^2}}, \tag{6.61}$$

where $\tau \neq 0$, and the generalised class of time–frequency representations in Eq. (6.57) can be expressed as

$$C_z(t, f) = \int_{-\infty}^{\infty}\int_{-\infty}^{\infty} \frac{\sqrt{2\pi}\sigma}{\tau} \mathrm{e}^{-\frac{2\pi^2\sigma^2 (t-t')^2}{\tau^2}} z\left(t' + \frac{\tau}{2}\right) z^*\left(t' - \frac{\tau}{2}\right) \mathrm{e}^{-\mathrm{i}2\pi f\tau} \,\mathrm{d}t'\mathrm{d}\tau. \tag{6.62}$$

This generalised class of time–frequency representations with exponential kernels preserves the signal energy marginals and can suppress cross-term interference substantially, with little broadening in the auto-term WVDs (Jones & Parks, 1992).

If the generalised kernel $K(t, f)$ is the smoothing filter of Eq. (6.43), and is separable into time and frequency components, the generalised class of time–frequency representations can be expressed as the smoothed WVD, $C_z(t, f) = \tilde{S}_{\mathrm{WVD}}(t, f)$. Because of the independence of the frequency- and time-axis filters, these two filters $g(t)$ and $\hat{h}(f)$ can be adjusted separately to a desired level of cross-term interference suppression (Claasen & Mecklenbräuker, 1980a, 1980b, 1980c, 1981; Auger & Flandrin, 1995; Baydar & Ball, 2001). The ability to remove the cross-term interference depends on the breadths of the two window functions, defined by σ_t and σ_f, respectively, and the smoothed WVD can achieve non-negativity, $\tilde{S}_{\mathrm{WVD}}(t, f) \geq 0$.

6.6 The Ambiguity Function and the Generalised WVD

The convolution form of the generalised autocorrelation $R(t,\tau)$ of Eq. (6.56) can be expressed equivalently as a multiplication in the frequency domain. Then, the generalised class of time–frequency representations (Eq. 6.57) may also be written as (Cohen, 1966)

$$C_z(t,f) = \int_{-\infty}^{\infty}\int_{-\infty}^{\infty} \hat{\gamma}(\xi,\tau)A_z(\xi,\tau)\mathrm{e}^{\mathrm{i}2\pi(\xi t - f\tau)}\mathrm{d}\tau\mathrm{d}\xi, \tag{6.63}$$

where $A_z(\xi,\tau)$ is called the ambiguity function, that is defined as

$$A_z(\xi,\tau) = \int_{-\infty}^{\infty} z\left(t+\frac{\tau}{2}\right)z^*\left(t-\frac{\tau}{2}\right)\mathrm{e}^{-\mathrm{i}2\pi\xi t}\mathrm{d}t. \tag{6.64}$$

The ambiguity function $A_z(\xi,\tau)$ is a Fourier transform with respect to time t. This form contrasts with the WVD, which is the Fourier transform with respect to the autocorrelation time lag τ.

The ambiguity function $A_z(\xi,\tau)$ is a signature of the signal that is invariant under time and frequency translations. The kernel function $\hat{\gamma}(\xi,\tau)$ works as a weighting function that is applied to $A_z(\xi,\tau)$.

The ambiguity function can also be defined as an inverse Fourier transform of the frequency-domain instantaneous autocorrelation:

$$A_z(\xi,\tau) = \int_{-\infty}^{\infty} \hat{z}\left(f+\frac{\xi}{2}\right)\hat{z}^*\left(f-\frac{\xi}{2}\right)\mathrm{e}^{\mathrm{i}2\pi f\tau}\mathrm{d}f. \tag{6.65}$$

Setting $\hat{\gamma}(\xi,\tau)=1$, Eq. (6.63) reveals that the ambiguity function $A_z(\xi,\tau)$ is a two-dimensional Fourier transform of $S_{\mathrm{WVD}}(t,f) = C_z(t,f)$:

$$A_z(\xi,\tau) = \int_{-\infty}^{\infty}\int_{-\infty}^{\infty} S_{\mathrm{WVD}}(t,f)\mathrm{e}^{-\mathrm{i}2\pi(\xi t - f\tau)}\mathrm{d}f\mathrm{d}t. \tag{6.66}$$

This relationship between the ambiguity function and the WVD reveals that the variable τ is the 'delay time' of the signal, as presented in the factor $\mathrm{e}^{\mathrm{i}2\pi f\tau}$, and the variable ξ is called the 'doppler frequency', as presented in the factor $\mathrm{e}^{-\mathrm{i}2\pi\xi t}$. Both variables in the ambiguity function $A_z(\xi,\tau)$ are not the absolute values; rather, they are relative shift values.

Equation (6.66) can be verified by inserting the WVD definition of Eq. (6.12) into the integrand, as

$$\begin{aligned}
&\int_{-\infty}^{\infty}\int_{-\infty}^{\infty} S_{\mathrm{WVD}}(t,f)\mathrm{e}^{-\mathrm{i}2\pi(\xi t - f\tau)}\mathrm{d}f\mathrm{d}t \\
&= \int_{-\infty}^{\infty}\int_{-\infty}^{\infty} \mathrm{e}^{-\mathrm{i}2\pi(\xi t - f\tau)}\mathrm{d}f\mathrm{d}t \int_{-\infty}^{\infty} z\left(t+\frac{\tau'}{2}\right)z^*\left(t-\frac{\tau'}{2}\right)\mathrm{e}^{-\mathrm{i}2\pi f\tau'}\mathrm{d}\tau' \qquad \text{(cont.)}
\end{aligned}$$

$$= \int_{-\infty}^{\infty} e^{-i2\pi\xi t} dt \int_{-\infty}^{\infty} \hat{z}\left(t+\frac{f}{2}\right)\hat{z}^*\left(t-\frac{f}{2}\right) e^{i2\pi f\tau} df$$

$$= \int_{-\infty}^{\infty} z\left(t+\frac{\tau}{2}\right) z^*\left(t-\frac{\tau}{2}\right) e^{-i2\pi\xi t} dt = A_z(\xi,\tau). \tag{6.67}$$

Since $A_z(\xi,\tau)$ is the Fourier transform of $S_{\mathrm{WVD}}(t,f)$, by substituting the ambiguity function of Eq. (6.64) back into the generalised class of representations of Eq. (6.63), we can derive directly the generalised class of representations $C_z(t,f)$ in Eq. (6.59).

In addition to the time- and frequency-domain formulae, the ambiguity function $A_z(\xi,\tau)$ may also be expressed as

$$A_z(\xi,\tau) = e^{i\pi\xi\tau} \int_{-\infty}^{\infty}\int_{-\infty}^{\infty} \rho_z(t,f) e^{-i2\pi(\xi t - f\tau)} dt df, \tag{6.68}$$

where $\rho_z(t,f)$ is the complex energy density function,

$$\rho_z(t,f) = z(t)\hat{z}^*(f) e^{-i2\pi ft}. \tag{6.69}$$

This complex function is called the Rihaczek distribution, which provides not only a time-varying energy spectrum but also a phase spectrum, $\arg\{\rho_z(t,f)\} = \arg\{z(t)\} - \arg\{\hat{z}(f)\} - 2\pi ft$, with strong time and frequency support properties (Levin, 1964; Rihaczek, 1968).

The formula of Eq. (6.68) can be proved as follows:

$$A_z(\xi,\tau) = e^{i\pi\xi\tau} \int_{-\infty}^{\infty}\int_{-\infty}^{\infty} z(t)\hat{z}^*(f) e^{-i2\pi ft} e^{-i2\pi(\xi t - f\tau)} dt df$$

$$= \int_{-\infty}^{\infty} \hat{z}^*(f) e^{i2\pi(f+\xi/2)\tau} df \int_{-\infty}^{\infty} z(t) e^{-i2\pi(f+\xi)t} dt$$

$$= \int_{-\infty}^{\infty} \hat{z}(f+\xi)\hat{z}^*(f) e^{i2\pi(f+\xi/2)\tau} df = \int_{-\infty}^{\infty} \hat{z}\left(f'+\frac{\xi}{2}\right)\hat{z}^*\left(f'-\frac{\xi}{2}\right) e^{i2\pi f'\tau} df', \tag{6.70}$$

which is exactly Eq. (6.65).

Given the complex energy density function $\rho_z(t,f)$, it might be easy to understand the physical meaning of the ambiguity function $A_z(\xi,\tau)$. The two-dimensional autocorrelation of $\rho_z(t,f)$ leads to the power (squared magnitude) of the ambiguity function (Stutt, 1964):

$$\int_{-\infty}^{\infty}\int_{-\infty}^{\infty} \rho_z(t,f)\rho_z^*(t+\tau, f+\xi)\mathrm{d}t\mathrm{d}f = \left|A_z(\xi,\tau)\right|^2. \tag{6.71}$$

Therefore, the ambiguity function $A_z(\xi,\tau)$ represents the resolution of the two-dimensional autocorrelation of the complex energy density function $\rho_z(t,f)$ in the joint time–frequency domain.

The ambiguity function, like the WVD, measures the time–frequency correlation of a signal, that is, the degree of similarity between $z(t)$ and its shifted-plus-modulated version. An important property of the ambiguity function relates to the interference geometry. The elements of the ambiguity function corresponding to the signal components are located in the vicinity of the origin, whereas those of the interferences are located far away from the origin. The distance of these terms from the origin is proportional to the time–frequency distance between the components that are involved.

For the signal that is defined in Eq. (6.40), $z(t) = c_m z_m(t) + c_n z_n(t)$, which is composed of two Gabor wavelets, the two auto-terms in the ambiguity function can be derived as

$$\begin{aligned} A_{z_m}(\xi,\tau) &= \sqrt{\pi}\,\sigma \mathrm{e}^{-\frac{\tau^2}{4\sigma^2}-\pi^2\sigma^2\xi^2}\mathrm{e}^{\mathrm{i}2\pi(f_m\tau-\xi t_m)}, \\ A_{z_n}(\xi,\tau) &= \sqrt{\pi}\,\sigma \mathrm{e}^{-\frac{\tau^2}{4\sigma^2}-\pi^2\sigma^2\xi^2}\mathrm{e}^{\mathrm{i}2\pi(f_n\tau-\xi t_n)}. \end{aligned} \tag{6.72}$$

The two cross-terms in the ambiguity function can be derived as

$$\begin{aligned} A_{z_m,z_n}(\xi,\tau) &= \int_{-\infty}^{\infty} z_m\left(t+\frac{\tau}{2}\right) z_n^*\left(t-\frac{\tau}{2}\right)\mathrm{e}^{-\mathrm{i}2\pi\xi t}\mathrm{d}t \\ &= \sqrt{\pi}\,\sigma \mathrm{e}^{-\frac{(\tau-\Delta t)^2}{4\sigma^2}-\pi^2\sigma^2(\xi-\Delta f)^2}\mathrm{e}^{\mathrm{i}2\pi(f_0\tau-\xi t_0)}\mathrm{e}^{\mathrm{i}(2\pi t_0\Delta f+\Delta\theta)}, \\ A_{z_n,z_m}(\xi,\tau) &= \sqrt{\pi}\,\sigma \mathrm{e}^{-\frac{(\tau+\Delta t)^2}{4\sigma^2}-\pi^2\sigma^2(\xi+\Delta f)^2}\mathrm{e}^{\mathrm{i}2\pi(f_0\tau-\xi t_0)}\mathrm{e}^{-\mathrm{i}(2\pi t_0\Delta f+\Delta\theta)}. \end{aligned} \tag{6.73}$$

Figure 6.6 shows the real part of the ambiguity function of this example signal which, as previously mentioned, is composed of two Gabor wavelets. The auto- and cross-terms clearly show three features:

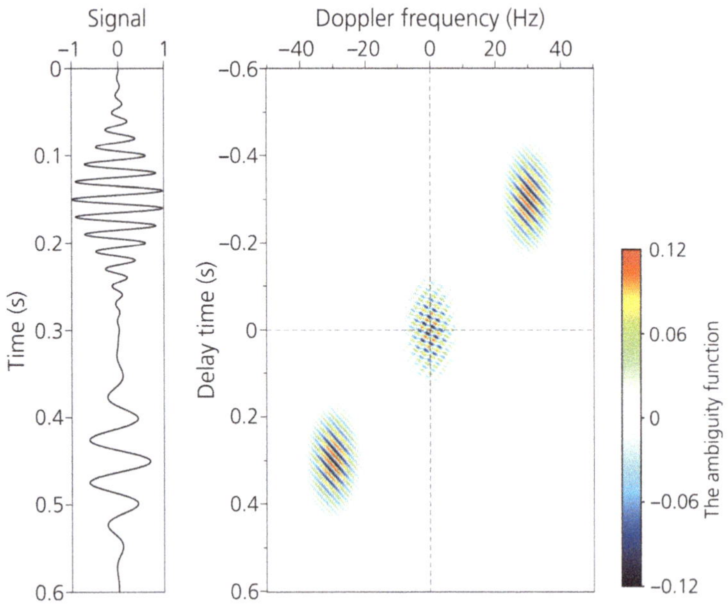

Figure 6.6 An example signal, consistin of two Gabor wavelets, and the ambiguity function (the real part). All auto-terms are localised at the origin, and the two cross-terms are positioned at $(\xi, \tau) = (\pm\Delta f, \pm\Delta t)$, respectively.

(1) The positions of any auto-terms $A_{z_m}(\xi, \tau)$ and $A_{z_n}(\xi, \tau)$ of the ambiguity function are independent of the dominant frequencies (f_m, f_n) and the central times (t_m, t_n).
(2) All auto-terms are localised at the origin $(\xi, \tau) = (0, 0)$.
(3) The two cross-terms are positioned at $(\xi, \tau) = (\pm\Delta f, \pm\Delta t)$, where $\Delta f = f_m - f_n$ and $\Delta t = t_m - t_n$.

The auto-terms are the signal terms. Recall that in the WVD the signal terms are localised in the respective time–frequency positions. Here in the ambiguity function, the signal terms are localised around the origin $(\xi, \tau) = (0, 0)$, and the cross-terms are localised away from the origin. This is an attractive feature which can be exploited for suppressing the cross-term interference in the WVD.

Because the generalised class of representations can be thought of as a generalised WVD with time- and frequency-translation invariance, and the WVD is the Fourier transform of the ambiguity function, the generalised class of representations can be treated as the Fourier transform of a generalised ambiguity function, with certain restrictions imposed. Introducing the generalised ambiguity function,

$$A_z^{\hat{\gamma}}(\xi,\tau)=\hat{\gamma}(\xi,\tau)A_z(\xi,\tau), \tag{6.74}$$

the generalised class of representations can be obtained by

$$C_z(t,f)=\int_{-\infty}^{\infty}\int_{-\infty}^{\infty}A_z^{\hat{\gamma}}(\xi,\tau)e^{i2\pi(\xi t-f\tau)}d\tau d\xi, \tag{6.75}$$

where the operator $e^{i2\pi\xi t}$ represents an inverse Fourier transform with respect to the 'doppler frequency' ξ, and the operator $e^{-i2\pi f\tau}$ represents a forward Fourier transform with respect to the 'delay time' τ. This two-dimensional Fourier transform generates the generalised WVD, $S_{\mathrm{WVD}}^{\gamma}(t,f)=C_z(t,f)$.

The relationship between the generalised ambiguity function and the generalised WVD suggests an alternate procedure for suppressing the cross-term interference existing in the WVD. The procedure consists of applying a low-pass filter $\hat{\gamma}(\xi,\tau)$ to the amplitude of the ambiguity function $A_z(\xi,\tau)$ and then applying the Fourier transform to the generalised ambiguity function $A_z^{\hat{\gamma}}(\xi,\tau)$, in order to produce a desirable generalised WVD, on which the signal complements are appropriately positioned.

For the example signal $z(t)=c_m z_m(t)+c_n z_n(t)$, composed of two Gabor wavelets, the generalised ambiguity function is

$$A_z^{\gamma}(\xi,\tau)=A_{z_m}(\xi,\tau)+A_{z_n}(\xi,\tau), \tag{6.76}$$

given by Eq. (6.72). The generalised WVD, $S_{\mathrm{WVD}}^{\gamma}(t,f)=C_z(t,f)$, can be obtained, following Eq. (6.75), by using a two-dimensional Fourier transform as

$$S_{\mathrm{WVD}}^{\gamma}(t,f)=P_{z_m}(t,f)+P_{z_n}(t,f), \tag{6.77}$$

where

$$\begin{aligned} P_{z_m}(t,f)&=2\sqrt{\pi}\sigma e^{-\frac{(t-t_m)^2}{\sigma^2}-4\pi^2\sigma^2(f-f_m)^2},\\ P_{z_n}(t,f)&=2\sqrt{\pi}\sigma e^{-\frac{(t-t_n)^2}{\sigma^2}-4\pi^2\sigma^2(f-f_n)^2}. \end{aligned} \tag{6.78}$$

These are exactly the WVD given by Eq. (6.39), for a Gabor wavelet.

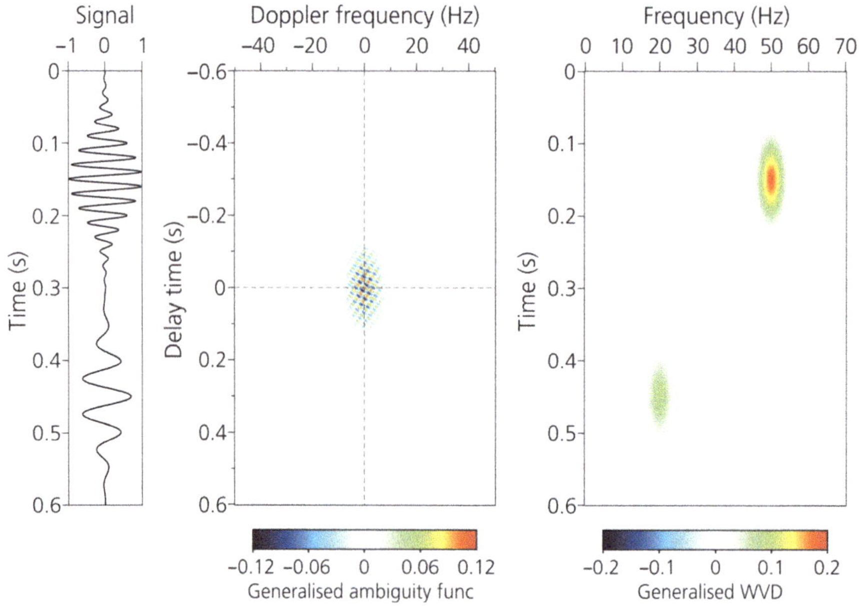

Figure 6.7 The generalised ambiguity function and the generalised WVD. The generalised ambiguity function keeps all auto-terms localised at the origin, and the generalised WVD keeps the auto-terms of the WVD at their respective positions.

The generalised ambiguity function is complex, and Figure 6.7 shows only the real part. But the generalised WVD that is produced by the two-dimensional Fourier transform is real. The positivity of the resultant generalised WVD, $S^{\gamma}_{\mathrm{WVD}}(t,f) \geq 0$, can be imposed as a restriction when designing the parameterisation kernel $\hat{\gamma}(\xi,\tau)$ for the generalised ambiguity function $A^{\hat{\gamma}}_{z}(\xi,\tau) = \hat{\gamma}(\xi,\tau)A_{z}(\xi,\tau)$.

Recall that the smoothed WVD $\tilde{S}_{\mathrm{WVD}}(t,f)$ employs a two-dimensional smoothing filter, and smoothing inevitably reduces the resolution, in both time and frequency. However, the generalised WVD $S^{\gamma}_{\mathrm{WVD}}(t,f)$ produces an accurate WVD energy density distribution, which has a better resolution than the smoothed WVD.

6.7 Implementation of the Standard and Smoothed WVDs

For the standard WVD, theoretically, the time-domain definition and the frequency-domain definition are equivalent, but their implementations have different accuracies in the time–frequency representations.

The time-domain WVD definition in Eq. (6.12) can be implemented as to generate the frequency spectra at specified time positions sequentially. Substituting $u = t + \tau/2$, Eq. (6.12) can be reformulated as

$$S_{\text{WVD}}(t, f) = 2e^{i2\pi(2f)t} \int_{-\infty}^{\infty} z(u) z^*(2t - u) e^{-i2\pi(2f)u} du, \tag{6.79}$$

where $e^{i2\pi(2f)t}$ is the effect of the time position t. Therefore, the WVD is implemented in four steps:

(1) resampling the analytic signal $z(t) \rightarrow z(u)$, with half of the original time interval $\Delta u = \Delta t/2$;
(2) calculating the instantaneous autocorrelation $\phi(u,t) = z(u) z^*(2t - u)$, at any specified time t;
(3) performing the Fourier transform $\hat{\phi}(2f,t) = \mathcal{F}_u[\phi(u,t)]$;
(4) calculating the WVD by $S_{\text{WVD}}(t, f) = 2e^{i2\pi(2f)t} \hat{\phi}(2f,t)$.

Steps 2 to 4 are repeated for all required time positions.

Resampling in step 1 is implemented by a Fourier transform pair: (i) performing the forward Fourier transform $\hat{z}(\xi) = \mathcal{F}_t[z(t)]$, in which the maximum frequency of the discretised analytic signal is $f_{\max} = 1/\Delta t$;

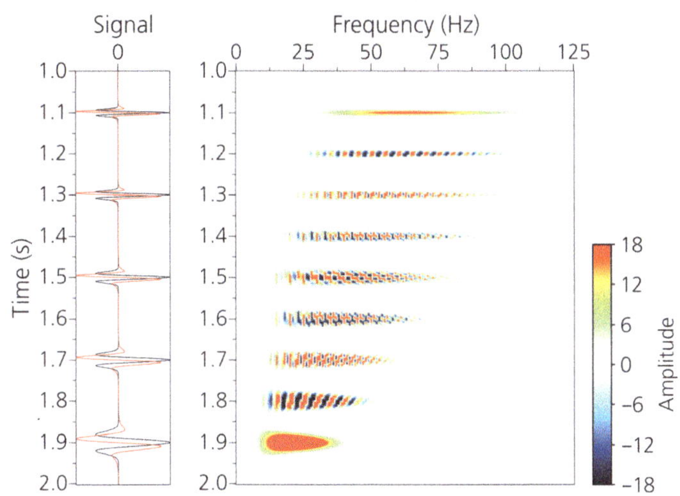

Figure 6.8 An analytic trace $z(t)$, formed by a synthetic trace (black curve) with five wavelets and the quadrature (red curve), and the WVD, implemented with the time-domain definition. The energy distribution is mixed with cross-term interference which has both positive and negative values.

(ii) padding the frequency-domain analytic signal $\hat{z}(\xi)$ with zeros within the range $(f_{max}, 2f_{max}]$; (iii) performing an inverse Fourier transform over the zero-padded analytic signal. The resampled analytic signal $z(u)$ is assumed to be nonperiodic in step 2, for the calculation of $\phi(u,t)$.

Figure 6.8 demonstrates a synthetic trace $x(t)$ (black) consisting of five wavelets, and its quadrature $y(t)$ (in red). These two real traces form the analytic trace $z(t)$. The time–frequency spectrum shows that only the first and last wavelets, at 1.1 s and 1.9 s, respectively, have been properly presented, since there is no negative energy representation in the time–frequency plane. The remaining three wavelets, at 1.3 s, 1.5 s, and 1.7 s, have their individual energy representations overlaid by the cross-term interference between any pair of wavelets. The cross-term interference has both positive and negative values.

The frequency-domain WVD definition in Eq. (6.13) can also be implemented as to generate the time spectra at specified frequencies sequentially. This implementation is attractive in practice, compared to the time-domain WVD implementation. Substituting $\eta = f + \xi/2$, Eq. (6.13) can be reformulated as

$$\tilde{S}_{WVD}(t,f) = 2e^{-i2\pi f(2t)} \int_{-\infty}^{\infty} \hat{z}(\eta)\hat{z}^*(2f-\eta)e^{i2\pi\eta(2t)}d\eta, \tag{6.80}$$

where $e^{-i2\pi f(2t)}$ is the frequency doppler.

The frequency-domain analytic signal $\hat{z}(\xi)$ needs to be resampled as $\hat{z}(\xi) \to \hat{z}(\eta)$ with a half of the frequency interval, $\Delta\eta = \Delta\xi/2$. Resampling is realised by time-domain zero-padding of the analytic signal $z(t)$, which makes $z(t)$ double the length on the time axis, prior to calculating the complex-to-complex Fourier transform, $\hat{z}(\eta) = \mathcal{F}_t[z(t)]$.

To calculate the frequency-domain instantaneous autocorrelation $\hat{\phi}(\eta, f) = \hat{z}(\eta)\hat{z}^*(2f-\eta)$, the analytic signal $\hat{z}(\eta)$ outside the range $[0, f_{max}]$ is taken based on the periodic property in the frequency domain.

Figure 6.9 displays the numerically calculated WVD $\tilde{S}_{WVD}(t,f)$, which is in complex, not in pure real value. This is caused by the implicit periodicity of the Fourier transform $\hat{z}(\xi) = \mathcal{F}_t[z(t)]$. It is evident in Figure 6.9, by way of the cross-term interference overlaying the real part (the upper panel) of the energy representation of the fifth wavelet at 1.9 s. This overlaid cross-term interference is caused by the ghost wavelets which might exist in the trace at time $t > 2$ s. The energy representation of the fifth wavelet in

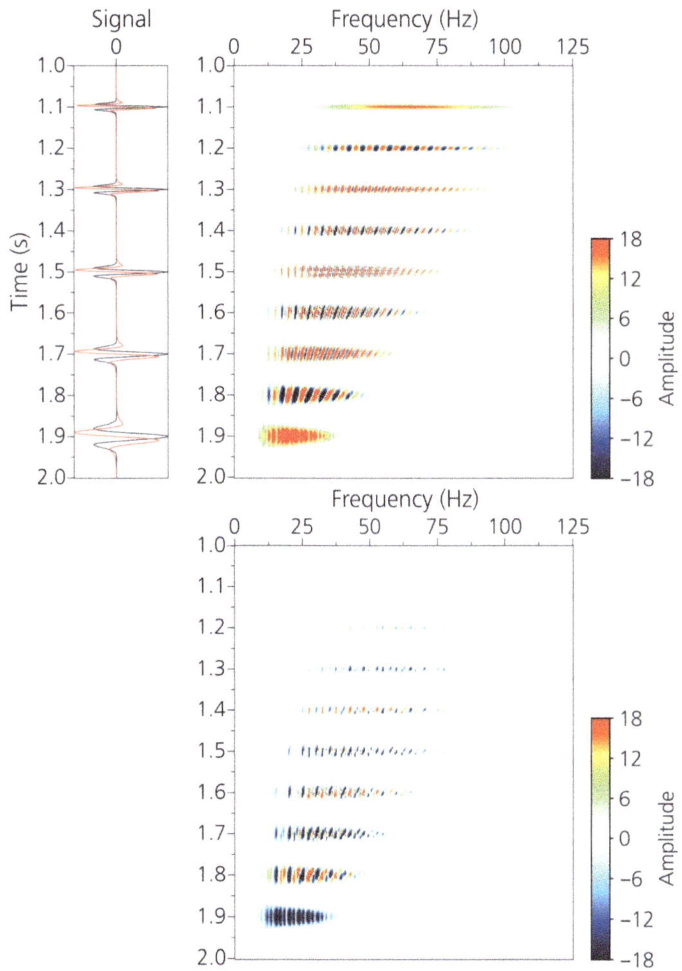

Figure 6.9 The analytic trace $z(t)$, and the WVD implemented with the definition in the frequency domain. Due to the implicit periodicity of the Fourier transform $\hat{z}(\xi) = \mathcal{F}_t[z(t)]$, the WVD is not in pure real values but has errors in the imaginary part (lower panel) of the complex values.

Figure 6.9 contrasts with that in Figure 6.8, in which the fifth wavelet is the last wavelet in the seismic trace, and the energy representation is not affected by any interference.

Because the WVD does not have phase information (Eq. 6.16), we can consider that the energy of the numerical WVD $\tilde{S}_{\mathrm{WVD}}(t, f)$ was 'split' into the real and the imaginary parts. Then, we can simply subtract the complex WVD

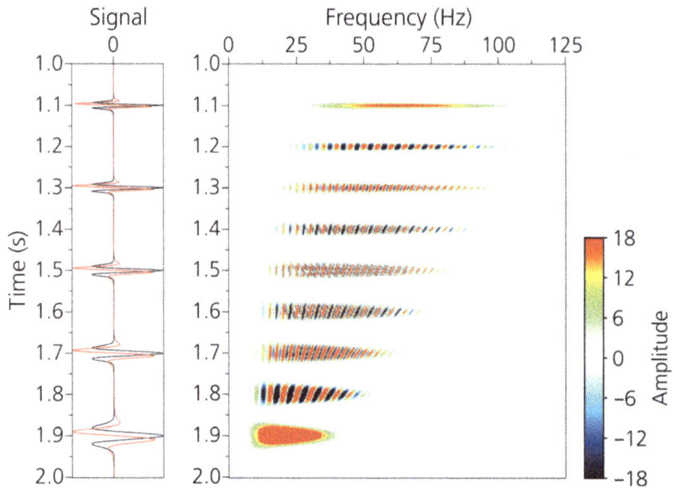

Figure 6.10 The WVD implemented using the WVD definition in the frequency domain, followed by a correction, in which the complex WVD spectrum is subtracted by its imaginary part.

$\tilde{S}_{\text{WVD}}(t,f)$ by the value of its imaginary part. Mathematically, the final WVD is obtained by a correction:

$$S_{\text{WVD}}(t,f) = \tilde{S}_{\text{WVD}}(t,f) - \text{Im}\{\tilde{S}_{\text{WVD}}(t,f)\}. \tag{6.81}$$

Of course, the imaginary part of $S_{\text{WVD}}(t,f)$ is zero after this subtraction. The final WVD, the real part of $S_{\text{WVD}}(t,f)$, is shown in Figure 6.10, which appears similar to the WVD (Figure 6.8) implemented using the time-domain WVD definition.

In summary, the WVD that is defined by the frequency-domain analytic signal $\hat{z}(\eta)$ is implemented in the following five steps:

(1) Performing the Fourier transform $\hat{z}(\eta) = \mathcal{F}_t[z(t)]$, in which the frequency interval $\Delta\eta = \Delta\xi/2$, and $\Delta\xi$ is the original interval of the analytic signal $\hat{z}(\xi)$ in the frequency domain.
(2) Calculating the frequency-domain instantaneous autocorrelation $\hat{\phi}(\eta,f) = \hat{z}(\eta)\hat{z}^*(2f-\eta)$, at any given frequency f.
(3) Performing an inverse Fourier transform $\phi(2t,f) = \mathcal{F}_\eta^{-1}\left[\hat{\phi}(\eta,f)\right]$.
(4) Calculating the WVD by $\tilde{S}_{\text{WVD}}(t,f) = 2e^{-i2\pi f(2t)}\phi(2t,f)$.
(5) Producing the final WVD result $|\tilde{S}_{\text{WVD}}(t,f)|$. based on $\tilde{S}_{\text{WVD}}(t,f)$.

Steps 2 to 5 are repeated for all required frequencies $f < f_{\max}$.

To smooth the WVD, the two-dimensional convolution of Eq. (6.44) can be implemented as a two-dimensional multiplication:

$$\tilde{S}_{\mathrm{WVD}}(t,f) = \int_{-\infty}^{\infty}\int_{-\infty}^{\infty} \hat{\gamma}(\xi,\tau) S_{\mathrm{WVD}}(\xi,\tau) \mathrm{e}^{-i2\pi(\xi t - f\tau)} \mathrm{d}\tau \mathrm{d}\xi, \tag{6.82}$$

where $\hat{\gamma}(\xi,\tau)$ is the two-dimensional Fourier transform of the smoothing filter $K(t,f)$, as defined in Eq. (6.58). For the smoothing filter $K(t,f)$ of Eq. (6.43), the Fourier transform is

$$\hat{\gamma}(\xi,\tau) = \mathrm{e}^{-2\pi^2\sigma_t^2\xi^2 - 2\pi^2\sigma_f^2\tau^2}. \tag{6.83}$$

Here, the frequency- and time-smoothing processes are separable.

The smoothing in the frequency axis can be implemented in the following three steps:

(1) Building the frequency-smoothing filter $h(\tau)$ in the time domain.
(2) Performing the Fourier transform of the WVD with respect to frequency f, $S_{\mathrm{WVD}}(t,\tau) = \mathcal{F}_f^{-1}[S_{\mathrm{WVD}}(t,f)]$, at all specified time positions t sequentially.
(3) Performing the inverse Fourier transform with respect to time τ, $\tilde{S}_{\mathrm{WVD}}(t,f) = \mathcal{F}_\tau[h(\tau)S_{\mathrm{WVD}}(t,\tau)]$, at all time positions t sequentially.

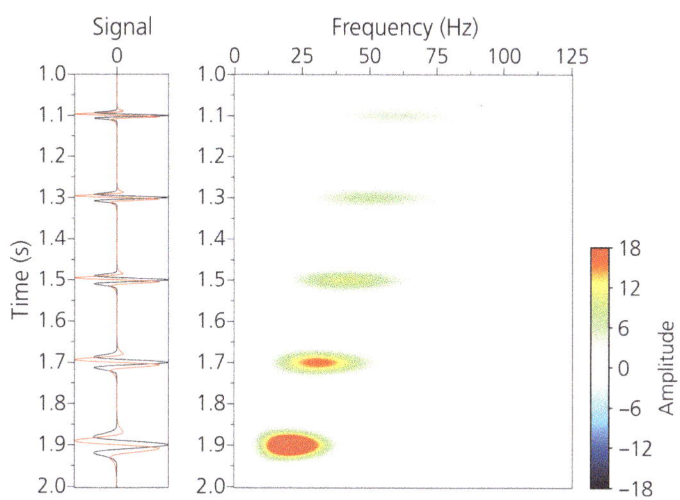

Figure 6.11 The smoothed WVD. The two-dimensional smoothing process can effectively suppress the cross-term interference.

The smoothing in the time axis can also be implemented in a similar way in three steps:

(1) Building the time-smoothing filter $\hat{g}(\xi)$ in the frequency domain.
(2) Performing the Fourier transform of the WVD $S_{\text{WVD}}(t, f)$ with respect to time t, $S_{\text{WVD}}(\xi, f) = \mathcal{F}_t[S_{\text{WVD}}(t, f)]$, at all frequency positions (f) sequentially.
(3) Performing the time smoothing via the Fourier transform, $\tilde{S}_{\text{WVD}}(t, f) = \mathcal{F}_\xi^{-1}[\hat{g}(\xi) S_{\text{WVD}}(\xi, f)]$, at all frequency positions (f) sequentially.

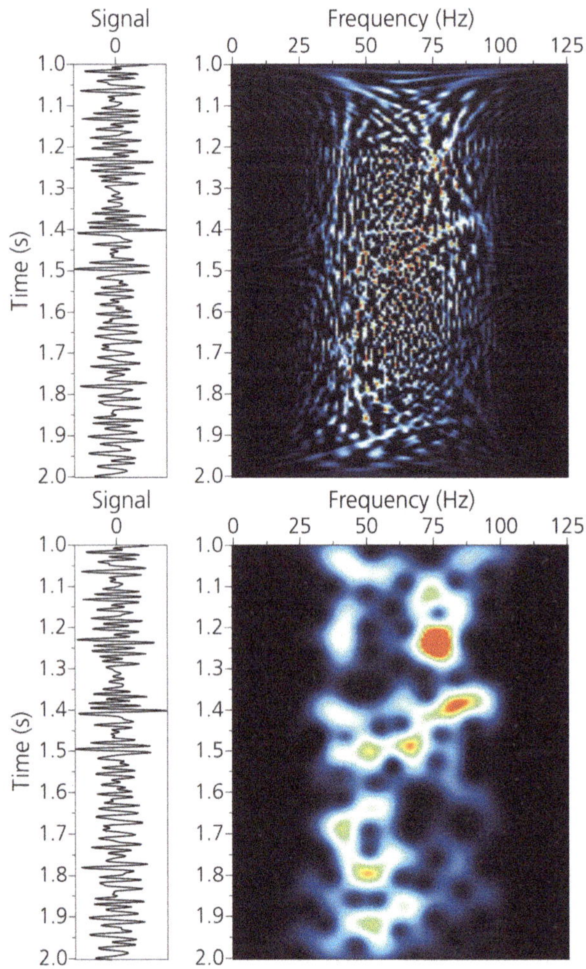

Figure 6.12 The example of a field seismic trace, the standard WVD $|S_{\text{WVD}}(t, f)|$, and the smoothed WVD $|\tilde{S}_{\text{WVD}}(t, f)|$.

Figure 6.11 demonstrates that the smoothing process can effectively suppress the cross-term interference. This is because the cross-term interference is featured as positive–negative value oscillatory, as shown in Figures 6.8 and 6.10. The smoothing process can cancel the negative values.

Figure 6.12 demonstrates the further use of a field seismic trace. The standard WVD $S_{\mathrm{WVD}}(t,f)$ has a very strong cross-term interference, and the spectrum cannot be used for interpretation. The smoothing process effectively suppresses the cross-term interference, and the smoothed WVD $\tilde{S}_{\mathrm{WVD}}(t,f)$ reveals the spectral features of the original seismic signal in the time–frequency plane.

6.8 Implementation of the Ambiguity Function and the Generalised WVD

The generalised WVD is realised through the ambiguity function. In the ambiguity function, the spectrum is completely insensitive to the time and frequency shifts in the signal. The ambiguity function represents these shifts as a phase factor.

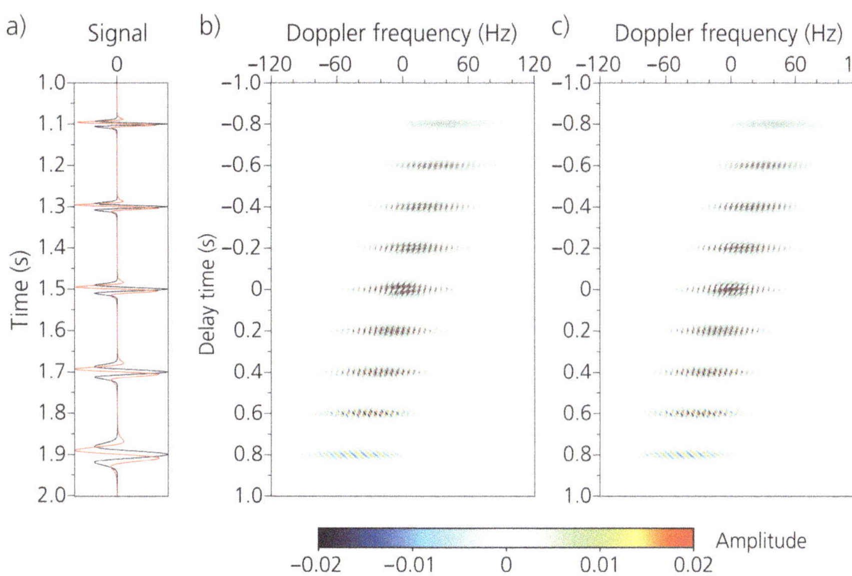

Figure 6.13 The ambiguity function $A_z(\xi,\tau)$, which is complex. (a) The analytic signal. (b) The real part of the ambiguity function. (c) The imaginary part of the ambiguity function.

A noteworthy difference between the ambiguity function and the WVD is that the WVD of any signal is real, while the ambiguity function is complex, as shown in Figure 6.13. The phase retains the information about the time and frequency shifts of the signal, even if the signal is real.

For the implementation, the ambiguity function $A_z(\xi,\tau)$ defined in Eq. (6.64) can be rewritten as

$$A_z(\xi,\tau) = e^{i\pi\xi\tau} \int_{-\infty}^{\infty} z(u) z^*(u-\tau) e^{-i2\pi\xi u} du. \tag{6.84}$$

A generalised ambiguity function $A_z^{\hat{\gamma}}(\xi,\tau)$ is a filtered version of the original ambiguity function (Eq. 6.74). The low-pass filter $\hat{\gamma}(\xi,\tau)$ is applied to $A_z(\xi,\tau)$ to mute out the cross-term interference and preserve the energy of the signal components which is concentrated in the vicinity of the origin and, in this example, is within a delay time window $[-0.1\text{s}, 0.1\text{s}]$.

The low-pass filter $\hat{\gamma}(\xi,\tau)$ is applied to the modulus $|A_z(\xi,\tau)|$ and does not affect the phase in the concentration area. Then, according to Eq. (6.75), the Fourier transform of the generalised ambiguity function $A_z^{\hat{\gamma}}(\xi,\tau)$ will produce the generalised WVD, $S_{\text{WVD}}^{\gamma}(t,f) = C_z(t,f)$.

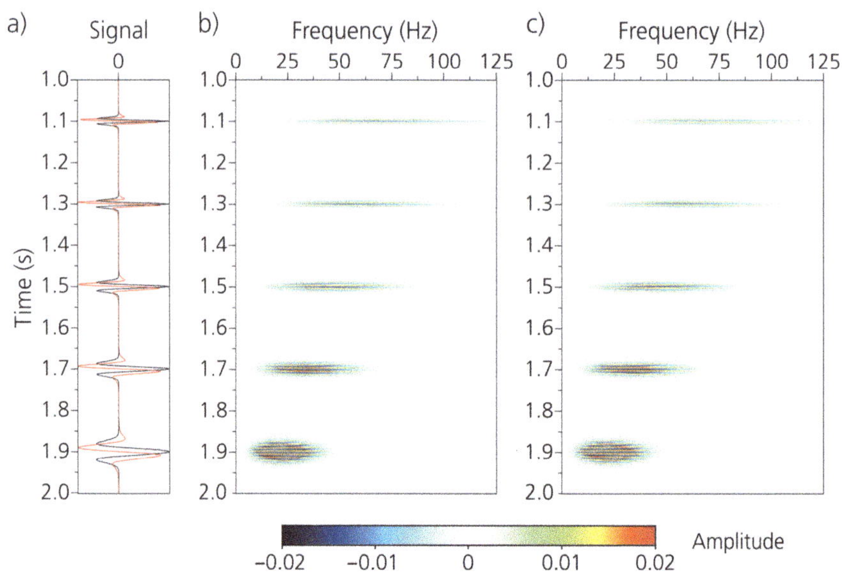

Figure 6.14 The generalised WVD. (a) The analytic signal $z(t)$. (b) The real part of the generalised WVD. (c) The imaginary part of the generalised WVD.

The inverse Fourier transform of $A_z^{\hat{\gamma}}(\xi,\tau)$ with respect to the 'doppler frequency' ξ is

$$A_z^{\gamma}(t,\tau)=\mathcal{F}_{\xi}^{-1}\left[A_z^{\hat{\gamma}}(\xi,\tau)\right]=e^{i2\pi f_{\mathrm{Nyq}}t}\mathcal{F}_{\eta}^{-1}\left[A_z^{\hat{\gamma}}(\eta,\tau)\right], \tag{6.85}$$

where $\xi=\eta-f_{\mathrm{Nyq}}$, $\xi\in\left[-f_{\mathrm{Nyq}},f_{\mathrm{Nyq}}\right]$, and f_{Nyq} is the Nyquist frequency. Then, the forward Fourier transform of $A_z^{\gamma}(t,\tau)$ with respect to the 'delay time' τ is

$$S_{\mathrm{WVD}}^{\gamma}(t,f)=\mathcal{F}_{\tau}\left[A_z^{\gamma}(t,\tau)\right]=e^{i2\pi f\tau_{\max}}\mathcal{F}_{u}\left[A_z^{\gamma}(t,u)\right], \tag{6.86}$$

where $\tau=u-\tau_{\max}$, and $\tau\in[-\tau_{\max},\tau_{\max}]$.

However, the value of the generalised WVD $S_{\mathrm{WVD}}^{\gamma}(t,f)$, as shown in Figure 6.14, is not real, and the imaginary part is not zero. This is because of the periodicity implicitly assumed in discrete Fourier transforms.

Figure 6.15 displays the final generalised WVD. This final time–frequency power spectrum is the modulus $\left|S_{\mathrm{WVD}}^{\gamma}(t,f)\right|$ of the complex WVD spectrum, since both the real part and the imaginary part of the generalised WVD $S_{\mathrm{WVD}}^{\gamma}(t,f)$ are the energy distribution.

Figure 6.16 displays the same field seismic trace example, along with the corresponding the ambiguity function $A_z(\xi,\tau)$ and the generalised WVD $S_{\mathrm{WVD}}^{\gamma}(t,f)$. Both $A_z(\xi,\tau)$ and $S_{\mathrm{WVD}}^{\gamma}(t,f)$ are complex, and for the sake of

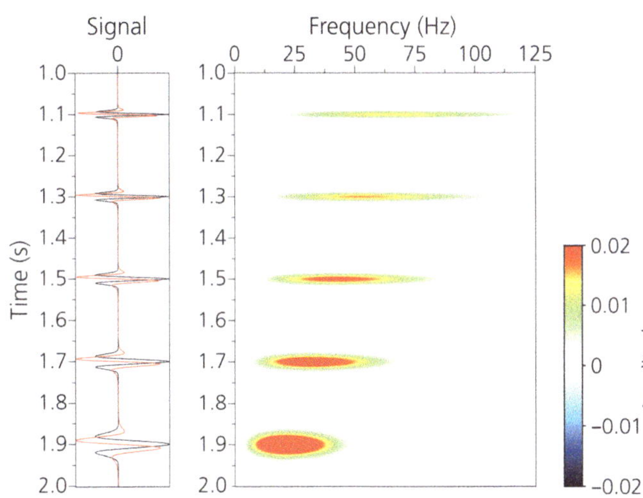

Figure 6.15 The analytic signal $z(t)$, and the final generalised WVD, which is the modulus $\left|S_{\mathrm{WVD}}^{\gamma}(t,f)\right|$.

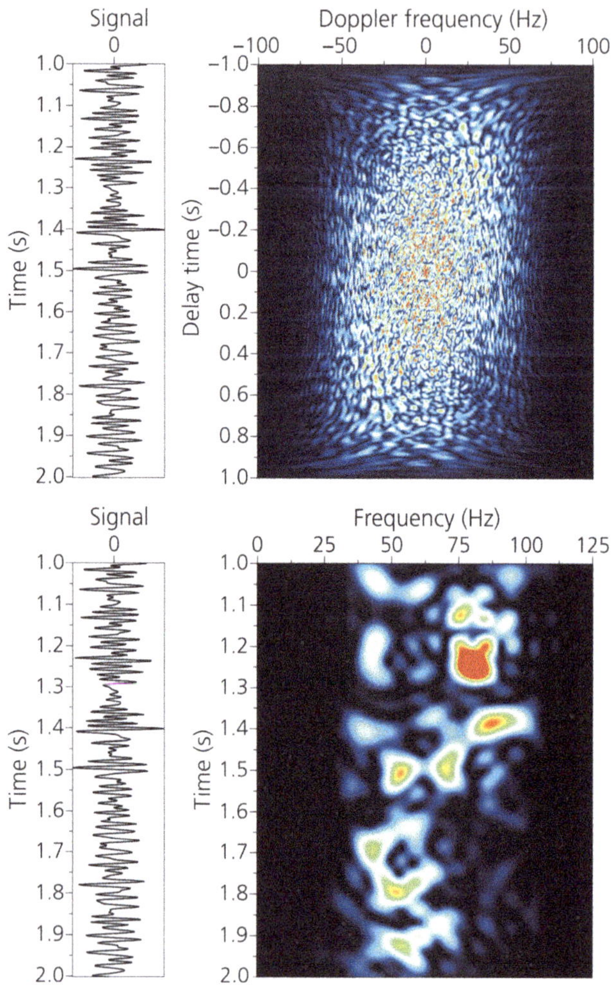

Figure 6.16 The field seismic trace, the ambiguity function $|A_z(\xi,\tau)|$, and the final generalised WVD $|S^{\gamma}_{\rm WVD}(t,f)|$.

visualisation, Figure 6.16 displays the modulus $|A_z(\xi,\tau)|$ and the modulus $|S^{\gamma}_{\rm WVD}(t,f)|$. Only a tiny portion of $A_z(\xi,\tau)$ in the vicinity of the origin is kept, for the reconstruction of the generalised WVD. The generalised WVD $|S^{\gamma}_{\rm WVD}(t,f)|$ (Figure 6.16) has a similar resolution to that of the smoothed WVD $\tilde{S}_{\rm WVD}(t,f)$ (Figure 6.12).

7

Matching Pursuit

The matching pursuit method decomposes a nonstationary seismic trace into a series of wavelets and then generates the energy density distribution in the time–frequency plane, using the Wigner–Ville distribution (WVD) method. Since the WVD of each wavelet is calculated separately, there is no cross-term interference in matching pursuit-based WVD, as is often the case with standard WVD.

Since the standard WVD spectrum represents the energy density distribution, the original seismic trace cannot be recovered from the WVD spectrum. However, with the matching pursuit, the seismic trace is decomposed into wavelets before the WVD calculation, and the seismic trace can be linearly recovered by summing the decomposed wavelets.

Therefore, in addition to time–frequency analysis, matching pursuit can also be used directly in signal processing, for example, to compensate the decomposed wavelets for the anelastic effect in the subsurface media before reconstructing a processed seismic trace. Matching pursuit combines time–frequency analysis and signal processing for reliable geophysical characterisation and geological interpretation.

7.1 Basics of Matching Pursuit

In matching pursuit, signal components are represented by wavelets that match their time–frequency signatures in the original seismic trace. Therefore, the matching pursuit overcomes the shortcoming of the Gabor transform type methods.

Time–Frequency Analysis of Seismic Signals, First Edition. Yanghua Wang.

The Gabor transform type methods, including the Gabor transform itself, the S transform, and the W transform, are basically window-based Fourier transforms. They are not able to describe seismic signatures of varying size, because the effective 'wavelets' have a constant scale σ that is proportional to the window size. Although the continuous wavelet transform decomposes a trace over time–frequency wavelets of varying scales, a dilated wavelet is built by restricting its frequency to be inversely proportional to the scale parameter. Since the frequency is a pseudo frequency, cross-correlation between the seismic trace and the dilated wavelet cannot provide an accurate estimate of the frequency content of a waveform whose Fourier transform is localised.

The standard matching pursuit is a resource-hungry procedure, in which the appropriate wavelets are iteratively selected from a large and redundant dictionary (Mallat & Zhang, 1993; Capobianco, 2003). I have made two improvements to the efficiency of this algorithm as follows: (a) the use of attributes from analytic signal analysis, and (b) the use of derived analytical expressions at different stages of the decomposition.

While using a set of parameters to represent a wavelet explicitly, a scale parameter σ can be introduced into the constituent wavelet (Wang, 2007). The scale parameter σ controls the frequency bandwidth and consequently the time breadth of the wavelet. However, this free parameter was not considered by a standard matching pursuit, in which the breadth of a wavelet was set as a function of the central frequency (Liu & Marfurt, 2007). Now, taking σ as a free parameter means adding an additional degree of freedom, which leads to a reduced decomposition residual. It also means, effectively, that many more wavelets in the dictionary are considered by the matching pursuit decomposition.

The matching pursuit adaptively decomposes a seismic trace into a series of constituent wavelets prior to WVD calculation, aiming to overcome the limitations of conventional methods for generating time–frequency spectra. Figure 7.1a displays a field seismic trace, and Figures 7.1b–d compare the time–frequency spectra produced by the Gabor transform, the continuous wavelet transform, and the matching pursuit, respectively. With the Gabor transform, the size of the time-window that slides along the trace is a predefined constant, and the time–frequency spectrum therefore depends on this predefined window size. In the continuous wavelet transform, the time duration of a constituent wavelet is also predefined, and the pseudo-frequency is inversely proportional to the time duration.

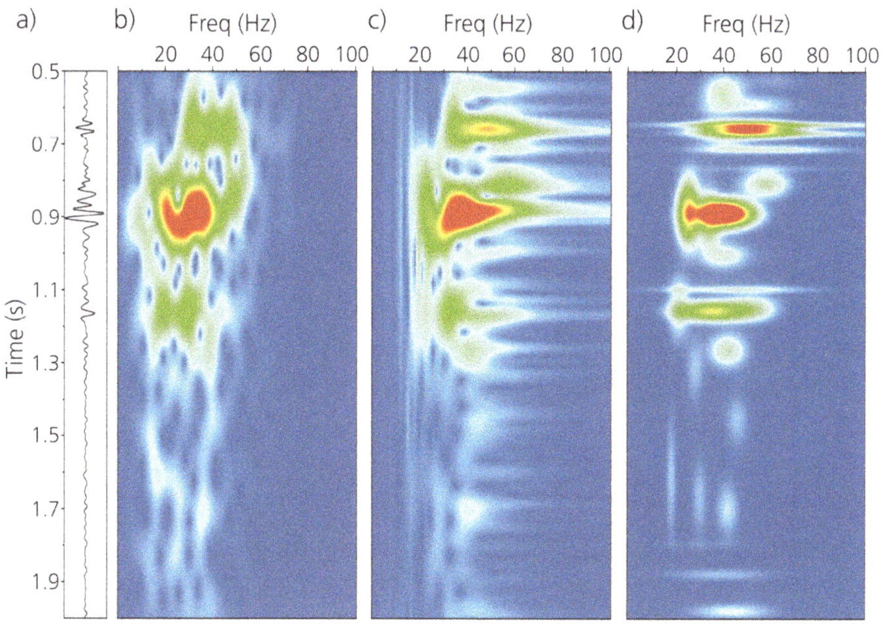

Figure 7.1 Time–frequency spectra generated by different methods. (a) A field seismic trace. (b) The spectrum produced by the Gabor transform. (c) The spectrum produced by the continuous wavelet transform. (d) The spectrum generated by the matching pursuit.

In contrast, the matching pursuit method uses a flexible wavelet size and can adaptively match the waveform signature in the nonstationary seismic trace. The resulting time–frequency spectrum shows that each wavelet has a distinct duration, on both the time and frequency axes.

7.2 Three-stage Matching Pursuit

In matching pursuit, a nonstationary seismic trace is decomposed over a set of basic wavelets, selected from a dictionary which is consisting of (1) Gabor wavelets, (2) wavelets augmented with a canonical basis, by discrete Dirac functions, and (3) the discrete Fourier basis of complex exponentials (Mallat & Zhang, 1993).

A Gabor wavelet is a complex function in the time domain (Section 3.1), and is represented as a complex sinusoid function that is multiplied by a Gaussian envelope (Gabor, 1946a, 1946b, 1946c):

$$w_z(t;\gamma) = g\left(\frac{t-u}{\sigma}\right)e^{i[2\pi f_c(t-u)+\theta]}, \tag{7.1}$$

where $w_z(t;\gamma)$ is the complex wavelet, $g(t)$ is a Gaussian function, u is the translation time, σ is the spread parameter in the time axis, f_c is the modulation frequency, and θ is the phase shift. Here, the complex Gabor wavelet also includes the phase shift θ, for an efficient evaluation. Thus, the Gabor wavelet is characterised by a set of four parameters, $\gamma = \{u, \sigma, f_c, \theta\}$.

For the decomposition of a real-valued seismic trace $x(t)$, a real-valued Gabor wavelet is used in matching pursuit:

$$w(t;\gamma) = g\left(\frac{t-u}{\sigma}\right)\cos\left[2\pi f_c(t-u)+\theta\right]. \tag{7.2}$$

Here, $w(t;\gamma) = \text{Re}[w_z(t;\gamma)]$, the real part of the complex wavelet $w_z(t;\gamma)$ defined in Eq. (7.1).

Matching pursuit decomposition is implemented iteratively, and each iteration adaptively extracts an optimal form of the basic wavelet, $w(t;\gamma_n)$, where n is the iteration number. After N iterations, the seismic trace $x(t)$ is decomposed linearly into the following form:

$$x(t) = \sum_{n=1}^{N} a_n w(t;\gamma_n) + R^{(N)}\{x(t)\}, \tag{7.3}$$

where a_n is the amplitude of the nth basic wavelet $w(t;\gamma_n)$, and $R^{(N)}\{x(t)\}$ is the residual after N iterations.

The nth iteration of decomposition consists of the following three stages:

(1) Estimating the four parameters $\gamma_n = \{u_n, \sigma_n, f_{c,n}, \theta_n\}$ efficiently but approximately.
(2) Updating these four parameters for an optimal basic wavelet $w(t;\gamma_n)$.
(3) Evaluating the amplitude a_n.

This three-stage procedure is repeated through all iterations.

Stage 1 and stage 2 estimate the four parameters γ_n of the basic wavelet, using the following equation:

$$w(t;\gamma_n) = \arg\max_{w(t;\gamma_n)\in D} \frac{\left|\left\langle w(t;\gamma_n), R^{(n-1)}\{x(t)\}\right\rangle\right|}{\left\|w(t;\gamma_n)\right\|}, \tag{7.4}$$

where $D = \{w(t;\gamma)\}$ is a comprehensive dictionary of the basic wavelets, $\langle \cdot, \cdot \rangle$ denotes the inner product of two functions, and $\|w(t;\gamma_n)\|$ normalises the basic wavelet $w(t;\gamma_n)$. The decomposition procedure starts with $R^{(0)}\{x(t)\} = x(t)$.

In stage 1, the analytic signal analysis is conducted first, to estimate three parameters $\{u_n, f_{c,n}, \theta_n\}$: the time of the maximum envelope is set to be the time delay u_n, the corresponding instantaneous frequency is set to be the centre frequency $f_{c,n}$, and the instantaneous phase is set to be the phase θ_n of the basic wavelet. Then, Eq. (7.4) is used to search for the optimal parameter σ_n over a group of pre-selected, uniformly distributed σ values, when the remaining three parameters $\{u_n, f_{c,n}, \theta_n\}$ are fixed. Taking σ_n as a variable in the basic wavelet is a powerful feature of this matching pursuit process (Wang, 2007).

In stage 2, the four parameters $\gamma_n = \{u_n, \sigma_n, f_{c,n}, \theta_n\}$ are optimised by searching within a sub-dictionary. Because the instantaneous attributes were estimated in stage 1 at a discrete sampling time, and σ_n was also a discrete value, these parameters need to be refined, and Eq. (7.4) is used again for searching the refined basic wavelet. The searching range around a parameter ξ is set as $[\xi - \Delta\xi,\ \xi + \Delta\xi]$, where Δu is the time sampling interval, $\Delta\sigma$ is the sampling interval for the spread parameter σ, Δf_c is the frequency sampling interval, and $\Delta\theta$ is 5 degrees. This stage of processing is computationally intensive.

In stage 3, the amplitude of the optimal wavelet is estimated by

$$a_n = \frac{\left\langle w(t;\gamma_n), R^{(n-1)}\{x(t)\} \right\rangle}{\|w(t;\gamma_n)\|^2}. \tag{7.5}$$

Here, the amplitude a_n can be either positive or negative. For the current nth decompositions, $R^{(n-1)}\{x(t)\} = a_n w(t;\gamma_n) + R^{(n)}\{x(t)\}$, where the basic wavelet $w(t;\gamma_n)$ is orthogonal to the residual $R^{(n)}\{x(t)\}$, we have

$$\left\| R^{(n-1)}\{x(t)\} \right\|^2 = \frac{\left| \left\langle w(t;\gamma_n), R^{(n-1)}\{x(t)\} \right\rangle \right|^2}{\|w(t;\gamma_n)\|^2} + \left\| R^{(n)}\{x(t)\} \right\|^2. \tag{7.6}$$

According to Eq. (7.4) for the orthogonal projection, a wavelet $w(t;\gamma_n)$ is chosen, such that the first term on the right-hand side of Eq. (7.6) is the maximum, and thus the second term $\left\| R^{(n)}\{x(t)\} \right\|$ is minimised. Therefore, Eq. (7.5) satisfies the minimum residual condition:

$$\{a_n, \gamma_n\} = \arg \min_{a_n, w(t;\gamma_n)} \left\| R^{(n)}\{x(t)\} \right\|^2 . \tag{7.7}$$

The final minimum residual $R^{(N)}\{x(t)\}$, after N iterations, is regarded as the data noise.

An improvement to the estimation of amplitude a_n can be made by, at each iteration, re-estimating the amplitudes of all previously extracted wavelets simultaneously in a least-squares sense, instead of calculating the amplitude of the currently extracted wavelet alone.

At the nth iteration, the amplitudes $\{a_1, a_2, \cdots, a_n\}$ of all n wavelets are re-estimated by minimising the following objective function:

$$\phi(a_1, a_2, \cdots, a_n) = \int_{-\infty}^{\infty} \left(x(t) - \sum_{i=1}^{n} a_i w(t;\gamma_i) \right)^2 \mathrm{d}t. \tag{7.8}$$

The least-squares solution indicates that all selected wavelets $\{a_i w(t;\gamma_i),\ i = 1, 2, \cdots, n\}$ are orthogonal to the residual $R^{(n)}\{x(t)\}$ at the nth iteration. This method is referred to as the orthogonal matching pursuit (Pati *et al.*, 1993; Rebollo-Neira & Lowe, 2002; Andrle *et al.*, 2004; Andrle & Rebollo-Neira, 2006; Tropp & Gilbert, 2007). However, the orthogonality here seems misleading, because the selected wavelets are not necessarily orthogonal to each other. Instead, I would emphasise that the simultaneous estimation of all amplitudes $\{a_i, i = 1, 2, \cdots, n\}$ is in a least-squares sense.

Once the seismic trace $x(t)$ is decomposed into a series of wavelets $a_n w(t;\gamma_n)$, for $n = \{1, 2, \cdots, N\}$, the energy density distribution in the time–frequency plane is generated, using the WVD method, by

$$E(t, f) = \sum_{n=1}^{N} \frac{a_n^2}{\left\| w(t;\gamma_n) \right\|^2} E_n(t, f), \tag{7.9}$$

where $E_n(t, f)$ is the WVD of the selected nth wavelet:

$$E_n(t, f) = \int_{-\infty}^{+\infty} w_{z,n}\left(t + \frac{\tau}{2}\right) w_{z,n}^*\left(t - \frac{\tau}{2}\right) \mathrm{e}^{-\mathrm{i}2\pi f \tau} \mathrm{d}\tau, \tag{7.10}$$

$w_{z,n}(t) = w_z(t;\gamma_n)$ is the analytic signal of the real-valued basic wavelet $w(t;\gamma_n)$, and $w_{z,n}^*(t)$ is the complex conjugate of $w_{z,n}(t)$.

To evaluate the normalisation factor $\left\| w(t;\gamma_n) \right\|^2$, the cosine function in the basic wavelet $w(t;\gamma_n)$ can be expanded into two complex exponentials.

Then, the module $\|w(t;\gamma_n)\|^2$ is formulated in terms of two integrals, as in the following:

$$\|w(t;\gamma_n)\|^2 = \frac{1}{2}\int_{-\infty}^{\infty} g^2\left(\frac{t-u_n}{\sigma_n}\right)\mathrm{d}t + \frac{1}{2}\mathrm{Re}\left(\mathrm{e}^{-\mathrm{i}2\theta_n}\int_{-\infty}^{\infty} g^2\left(\frac{t-u_n}{\sigma_n}\right)\mathrm{e}^{-\mathrm{i}2\pi(2f_{c,n})(t-u_n)}\mathrm{d}t\right). \tag{7.11}$$

This expression can be calculated efficiently through a fast Fourier transform (FFT), in which the amplitudes at frequency zero and $2f_{c,n}$ produce the first and second terms, respectively.

The W transform in Chapter 5 can also be used to generate the time–frequency spectrum (Luo *et al.*, 2022). Instead of evaluating the instantaneous frequency, matching pursuit provides a reliable dominant frequency for each wavelet that is decomposed from the seismic signal.

7.3 Matching Pursuit with the Morlet Wavelet

The Morlet wavelet is a complex Gabor wavelet that is suitable for the energy and frequency quantification of seismic signals, because the Morlet wavelet involves the possible phase delay of waveforms, due to velocity dispersion (Morlet *et al.*, 1982a, 1982b). It is particularly appropriate for attenuation and resolution studies of seismic waves propagating through subsurface anelastic media.

The Morlet wavelet $m(t)$, centred at the abscissa u, is defined as

$$m(t-u) = \mathrm{e}^{-\frac{(4\ln 2)f_c^2(t-u)^2}{\sigma^2}}\mathrm{e}^{\mathrm{i}[2\pi f_c(t-u)+\theta]}, \tag{7.12}$$

where f_c is the central frequency, and σ is the spread parameter controlling the wavelet breadth. Here, the energy normalisation factor is dropped out from $\psi_\sigma(t)$, which is presented in Eq. (3.15), since the amplitude a is evaluated explicitly in matching pursuit.

The coefficient of the exponent in the Gaussian window function is to ensure that the time-domain Morlet wavelet has a constant shape ratio as

$$\frac{T}{f_c^{-1}} = \frac{\sigma}{2}, \tag{7.13}$$

where T is half of the wavelet breadth, measured at half of the maximum amplitude of the wavelet envelope. Therefore, the physical meaning of parameter σ is the ratio of the breadth ($2T$) to the dominant period (f_c^{-1}) of the time-domain wavelet.

The Fourier transform of the Morlet wavelet $m(t-u)$ is

$$\hat{m}(f)=\sqrt{\frac{\pi}{\ln 2}}\frac{\sigma}{2f_c}\mathrm{e}^{-\frac{\pi^2\sigma^2(f-f_c)^2}{(4\ln 2)f_c^2}}\mathrm{e}^{-\mathrm{i}(2\pi fu-\theta)}, \tag{7.14}$$

where $f_c > 0$ is assumed. The Morlet wavelet in the frequency domain also has a constant shape ratio:

$$\frac{B}{f_c}=\frac{2\ln 2}{\pi\sigma}, \tag{7.15}$$

where B is half of the frequency bandwidth.

Given the Morlet wavelet defined by Eq. (7.12) as the nth basic wavelet, the WVD $E_n(t, f)$ of Eq. (7.10) is

$$E_n(t,f)=\sqrt{\frac{\pi}{2\ln 2}}\frac{\sigma}{f_c}\mathrm{e}^{-\frac{\pi^2\sigma^2(f-f_c)^2}{(2\ln 2)f_c^2}-\frac{(8\ln 2)f_c^2t^2}{\sigma^2}}. \tag{7.16}$$

The WVD is real, because the Morlet wavelet is an analytic wavelet.

In matching pursuit, a real-valued seismic trace is decomposed using a real-valued Morlet wavelet, which is in the form

$$m(t-u)=\mathrm{e}^{-\frac{(4\ln 2)f_c^2(t-u)^2}{\sigma^2}}\cos\left[2\pi f_c(t-u)+\theta\right]. \tag{7.17}$$

Correspondingly, the two integrals in $\|w(t;\gamma)\|^2$ (Eq. 7.11) are

$$\int_{-\infty}^{\infty} g^2\left(\frac{t-u}{\sigma}\right)\mathrm{d}t=\frac{1}{2}\sqrt{\frac{\pi}{2\ln 2}}\frac{\sigma}{f_c}, \tag{7.18}$$

$$\int_{-\infty}^{\infty} g^2\left(\frac{t-u}{\sigma}\right)\mathrm{e}^{-\mathrm{i}4\pi f_c(t-u)}\mathrm{d}t=\frac{1}{2}\sqrt{\frac{\pi}{2\ln 2}}\frac{\sigma}{f_c}\mathrm{e}^{-\frac{\pi^2\sigma^2}{2\ln 2}}. \tag{7.19}$$

Therefore, the modulus $\|w(t;\gamma)\|^2$ has the following expression:

$$\|w(t;\gamma)\|^2=\frac{1}{4}\sqrt{\frac{\pi}{2\ln 2}}\frac{\sigma}{f_c}\left(1+\mathrm{e}^{-\frac{\pi^2\sigma^2}{2\ln 2}}\cos(2\theta)\right). \tag{7.20}$$

This analytical expression leads to an efficient calculation for Eqs. (7.4) and (7.5), and also for the analytical expression of Eq. (7.9).

In summary, matching pursuit uses the Morlet wavelet as a basic wavelet, for reasons of effectiveness and efficiency (Wang, 2007): (a) The Morlet wavelet can represent the attenuation behaviour of seismic wave propagation; (b) The use of the Morlet wavelet saves search time, instead of the resource-intensive search of a redundant wavelet dictionary; (c) By using an analytic form of the wavelet, analytical expressions are derived for the calculation of the wavelet decomposition and the generation of the time-frequency spectrum.

7.4 The Sigma Filter

After matching pursuit, by removing wavelets with either very small or very large σ values, any spikes or sinusoid functions can be suppressed effectively from the time–frequency spectrum. This is referred to as sigma (σ) filtering.

Matching pursuit provides extremely flexible signal representation, with the signal properties explicitly given by the amplitude, time, scale, frequency,

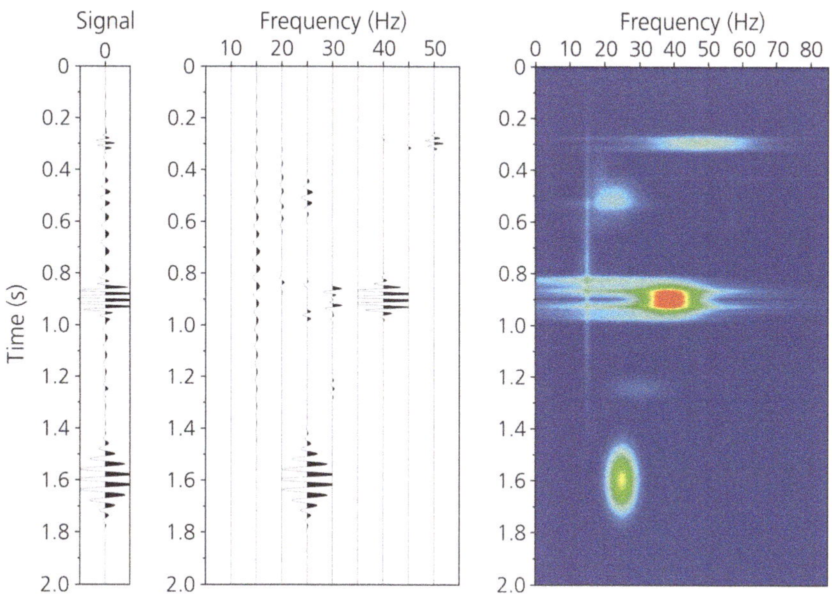

Figure 7.2 A synthetic seismic trace, the wavelets decomposed by matching pursuit, and the time–frequency representation of the energy density distribution.

and phase values of the decomposed wavelets. The synthetic example in Figure 7.2 shows that the wave package centred at about 0.3 s is, in fact, composed of three wavelets with mean frequencies around 40, 45, and 50 Hz, respectively. The wave package centred at about 0.9 s is also composed of several wavelets with different central frequencies, but the wave package centred at 1.6 s is indeed a single isolated wavelet.

For comparison, Figure 7.3 displays the time–frequency spectra generated by the Gabor transform. The input is the same synthetic trace of Figure 7.2. But Figure 7.3 shows that the time–frequency spectra strongly depend on the window size specified for the Gabor transform. In Figure 7.3, the spread parameter σ used in each panel from left to right is 17, 35, and 71, respectively.

In addition, the resolution of Figure 7.3 is lower than that in Figure 7.2, which is generated from matching pursuit. For example, the same wave package centred at 1.6 s appears differently in the time–frequency spectra with different window sizes. Between 0.4 and 0.6 s, the wavelet at 0.5 s and a 15 Hz sinusoid cannot be properly separated by the Gabor transform using either short or long window sizes.

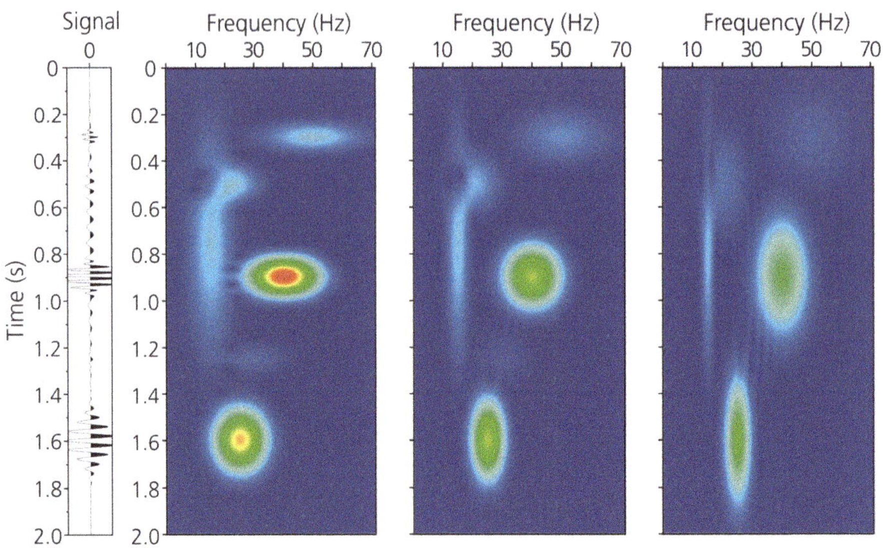

Figure 7.3 The Gabor transform with different window sizes. The spread parameter σ is, from left to right, 17, 35, and 71, respectively. The spectra produced by the Gabor transform are strongly dependent on the spread parameter σ, and the resolutions are generally lower than those of the matching pursuit.

In the matching pursuit scheme, the spread parameter σ, which controls the time breadth and the frequency band of a basic wavelet, is determined adaptively. Therefore, matching pursuit can extract even spikes and sinusoids. For example, the time–frequency spectrum in Figure 7.2 displays a sinusoid of 15 Hz within the 0.1–1.5 s time interval. Consequently, matching pursuit can separate neighbouring wavelets that are impossible to distinguish in the original seismic trace. The three wavelets around 0.3 s can be clearly identified from the matching pursuit-based time–frequency spectrum: one is an ellipse, and two are band-limited spikes.

The amplitude distribution of wavelets is visible as vertical lines for cosine functions, horizontal lines for spikes (Dirac δ), or ellipses with axes that are proportional to the time and frequency spread for the Gabor wavelets. Therefore, based on the time–frequency spectrum, one could design a filter that will reject certain wavelets, such as cosine functions, and spikes. This is the sigma filtering, to filter the decomposed wavelets or the time–frequency spectrum based on the σ value (Wang, 2007).

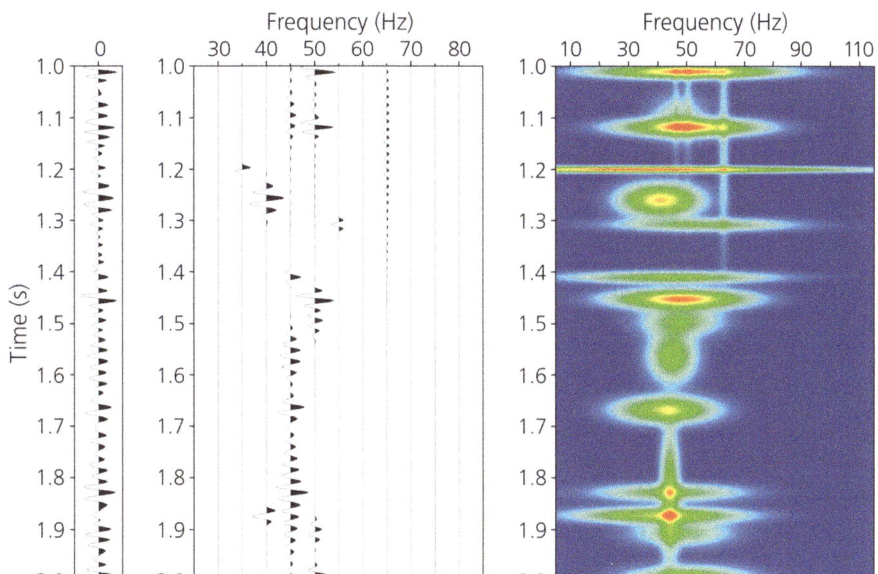

Figure 7.4 An example of a field seismic trace, decomposed wavelets with different central frequencies f_c, and the corresponding time–frequency representation. A central frequency value f_c on the top covers a range of central frequencies. For example, the $f_c = 50$ Hz trace consists of wavelets with central frequencies between 47.5 Hz and 52.5 Hz.

Figure 7.4 displays a field seismic trace that might be a good representation of land seismic data, decomposed wavelets, and the energy density distribution in the time–frequency plane. In the panel of decomposed wavelets, the central-frequency value f_c on the top axis is the centre value of a central-frequency range. For example, the $f_c = 50\,\text{Hz}$ trace consists of wavelets with central frequencies f_c between 47.5 and 52.5 Hz. Figure 7.4 shows that there are spikes and sinusoid functions.

For the Morlet wavelet $m(t)$ defined in Eq. (7.12), its Fourier transform $\hat{m}(f)$ is given by Eq. (7.14). The time and frequency extents of the wavelet can be estimated in terms of the spread parameter σ and the central frequency f_c. Half of the time window length can be estimated as

$$L_t = \frac{1}{4}\sqrt{\frac{\ln 10}{5\ln 2}|\varepsilon|}\,\frac{\sigma}{f_c} \approx 0.20\sqrt{|\varepsilon|}\frac{\sigma}{f_c}, \tag{7.21}$$

and half of the frequency window length is

$$L_f = \frac{f_c}{\pi\sigma}\sqrt{\frac{\ln 10 \cdot \ln 2}{5}|\varepsilon|} \approx 0.18\sqrt{|\varepsilon|}\frac{f_c}{\sigma}, \tag{7.22}$$

where ε is the minimum amplitude in dB, and is set as $\varepsilon = -120$ dB in the WVD calculation of each individual wavelet. Equations (7.21) and (7.22) indicate that the breadth in time and the bandwidth in frequency are inversely proportional to each other, and of course they depend on the spread parameter σ.

The property of σ dependency can be exploited to design a sigma (σ) filter. For a small σ value, the wavelet represents a spike that has a broad frequency band and a short duration in time. A large σ value represents a sinusoid function that has a narrow frequency band and a long duration in time.

In the field seismic trace example of Figure 7.4, by rejecting wavelets with $\sigma < 0.4$ and wavelets with $\sigma > 10$, we produce a result that is shown in Figure 7.5: the time–frequency spectrum and the corresponding wavelets with different central-frequencies f_c. Applying the sigma (σ) filter, the time–frequency spectrum can be cleaned up for geophysical analysis, such as gas reservoir detection. Moreover, a seismic trace can be reconstructed after rejecting unwanted wavelets using the sigma (σ) filter.

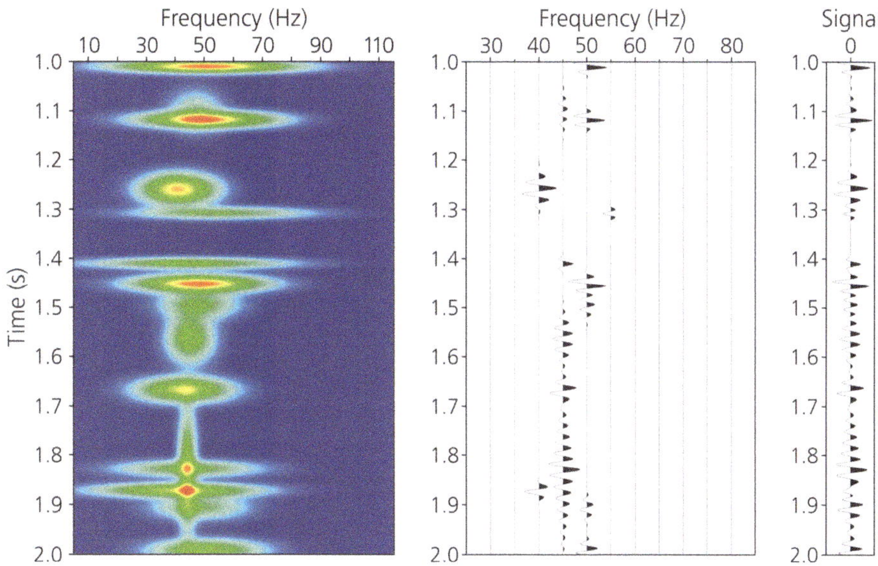

Figure 7.5 Sigma (σ) filtering. The time–frequency spectrum, corresponding wavelets with central frequencies in different ranges, and the reconstructed seismic trace after sigma filtering. In this example, the low cut-off value is $\sigma = 0.4$, and the high cut-off value is $\sigma = 10$.

7.5 Multichannel Matching Pursuit

Although matching pursuit can adaptively decompose a seismic trace into a series of wavelets, the solution is not unique, and it is also strongly influenced by data noise. The uniqueness of the matching pursuit can be improved by a multichannel implementation, exploiting lateral coherence as a constraint (Wang, 2010). The multichannel matching pursuit method (MCMP) extracts a basic wavelet that has optimal correlations with neighbouring seismic traces. According to linearity theory, a basic wavelet shared by adjacent traces best fits to the average of multiple traces. Therefore, MCMP can effectively suppress data noise and stabilise the decomposition.

Figure 7.6a displays a field seismic profile. Figure 7.6b reveals that, when applying a single channel matching pursuit to field seismic data, there is a non-uniqueness problem with the decomposition. Because of this non-uniqueness, the decomposition is also severely affected by data noise. As the single channel matching pursuit performs decomposition on each trace

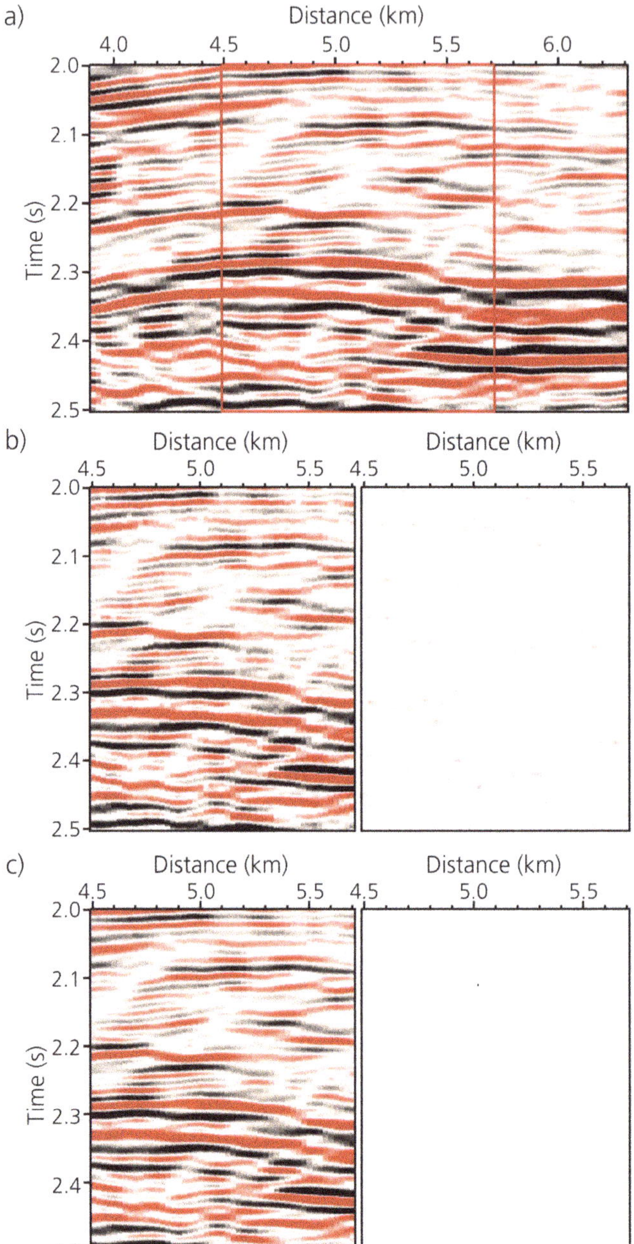

Figure 7.6 Completeness of multichannel matching pursuit. (a) A field seismic profile, on which the red frame indicates the zoom-in area shown in figures b and c. (b) Reconstructed profile by single channel matching pursuit and the residuals. (c) Reconstructed profile using multichannel matching pursuit and the residuals.

independently, the decomposed wavelets, and in turn the reconstructed profile (the left panel of Figure 7.6b), are lacking in spatial continuation between seismic traces, which results in some residuals (the right panel of Figure 7.6b). Consequently, the time–frequency spectrum will lack lateral continuation along a profile.

The multichannel implementation of matching pursuit may partially overcome the non-uniqueness problem (Figure 7.6c). With the constraint of lateral coherence, both the decomposed wavelets and the time–frequency spectrum will have optimal lateral continuation along the profile. Such constrained decomposition also stabilises the convergence of the iterative procedure, and, consequently, the reconstructed profile can accurately resemble the original seismic profile.

The core of the multichannel matching pursuit algorithm is that a basic wavelet is shared by adjacent traces. Assuming a group of neighbouring traces $\{x_1(t), x_2(t), \cdots, x_L(t)\}$ that would be around a single trace $x_\ell(t)$, the single trace $x_\ell(t)$ is decomposed into a series of wavelets, based on all L traces. At the beginning the residual trace is $R^{(0)}\{x_\ell(t)\} = x_\ell(t)$, and at the current nth decomposition, the residual trace is

$$R^{(n)}\{x_\ell(t)\} = R^{(n-1)}\{x_\ell(t)\} - a_{\ell,n} w(t;\gamma_n), \tag{7.23}$$

where $w(t;\gamma_n)$ is the basic wavelet that is extracted from the nth decomposition. Although $a_{\ell,n}$ is the amplitude of the basic wavelet $w(t;\gamma_n)$ for the individual trace $x_\ell(t)$, the basic wavelet $w(t;\gamma_n)$ is shared by all L traces within the group.

After N decompositions, the individual trace $x_\ell(t)$ is decomposed into N wavelets as

$$x_\ell(t) = \sum_{n=1}^{N} a_{\ell,n} w(t;\gamma_n) + R^{(N)}\{x_\ell(t)\}, \tag{7.24}$$

where $R^{(N)}\{x_\ell(t)\}$ is the final residual of the individual trace $x_\ell(t)$.

Within each decomposition, the implementation of multichannel matching pursuit can be divided into three steps:

(1) Performing a single channel matching pursuit to an average residual trace, to generate an initial estimate of the basic wavelet.
(2) Refining the basic wavelet in a multichannel fashion.
(3) Estimating the amplitudes of the wavelet corresponding to each individual trace.

In the first step, a single residual trace $R^{(n-1)}\{y(t)\}$ is generated by averaging over all L residual traces within the group:

$$R^{(n-1)}\{y(t)\} = \frac{1}{L}\sum_{\ell=1}^{L} R^{(n-1)}\{x_\ell(t)\}, \tag{7.25}$$

where $R^{(0)}\{y(t)\} = \frac{1}{L}\sum_{\ell=1}^{L} x_\ell(t)$. Then, a single channel matching pursuit is performed on the residual trace $R^{(n-1)}\{y(t)\}$, searching for the set of parameters $\gamma_n = \{u_n, \sigma_n, f_{c,n}, \theta_n\}$ to represent a basic wavelet $w(t;\gamma_n)$.

In the second step, the set of parameters $\gamma_n = \{u_n, \sigma_n, f_{c,n}, \theta_n\}$ is refined over a group of preselected, uniformly distributed values, by maximising the sum of the correlation coefficients in each residual trace:

$$w(t;\gamma_n) = \arg \max_{w(t;\gamma_n)\in D} \frac{\sum_{\ell=1}^{L}\left|\left\langle w(t;\gamma_n), R^{(n-1)}\{x_\ell(t)\}\right\rangle\right|}{\left\|w(t;\gamma_n)\right\|}. \tag{7.26}$$

Thus, the basic wavelet $w(t;\gamma_n)$ fits to all residual traces $R^{(n-1)}\{x_\ell(t)\}$ within the group.

In the third step, the amplitudes $a_{\ell,n}$ corresponding to each individual trace $\ell = \{1, 2, \cdots, L\}$ are estimated by

$$a_{\ell,n} = \frac{\left\langle w(t;\gamma_n), R^{(n-1)}\{x_\ell(t)\}\right\rangle}{\left\|w(t;\gamma_n)\right\|^2}. \tag{7.27}$$

The residual traces $R^{(n)}\{x_\ell(t)\}$ are given by Eq. (7.23).

These three steps are implemented iteratively for $n = \{1, 2, \cdots, N\}$. The procedure terminates when the residual energy is less than a threshold, or the number of decompositions reaches a preset maximum number.

The robust implementation of the multichannel matching pursuit scheme described above exploits the linearity of the residuum operator $R^{(n)}$ (Durka *et al.*, 2005).

Initially, since $R^{(0)}$ is an identity operator, $R^{(0)}\{y(t)\} = y(t)$, we have

$$\frac{1}{L}\sum_{\ell=1}^{L} R^{(0)}\{x_\ell(t)\} = R^{(0)}\left\{\frac{1}{L}\sum_{\ell=1}^{L} x_\ell(t)\right\} = R^{(0)}\{y(t)\}. \tag{7.28}$$

For the first iteration, the residuum operator $R^{(1)}$ is a linear operator such that

$$\begin{aligned}\frac{1}{L}\sum_{\ell=1}^{L} R^{(1)}\{x_\ell(t)\} &= \frac{1}{L}\sum_{\ell=1}^{L}\left[R^{(0)}\{x_\ell(t)\} - a_{\ell,1}w(t;\gamma_1)\right]\\ &= R^{(0)}\left\{\frac{1}{L}\sum_{\ell=1}^{L} x_\ell(t)\right\} - \frac{1}{L}\sum_{\ell=1}^{L} a_{\ell,1}w(t;\gamma_1)\\ &= R^{(1)}\left\{\frac{1}{L}\sum_{\ell=1}^{L} x_\ell(t)\right\} = R^{(1)}\{y(t)\}, \end{aligned} \tag{7.29}$$

where $\frac{1}{L}\sum_\ell a_{\ell,1}w(t;\gamma_1)$ is an average wavelet. By induction, for the nth iteration, $R^{(n)}$ is a linear operator so that

$$\begin{aligned}\frac{1}{L}\sum_{\ell=1}^{L} R^{(n)}\{x_\ell(t)\} &= \frac{1}{L}\sum_{\ell=1}^{L}\left[R^{(n-1)}\{x_\ell(t)\} - a_{\ell,n}w(t;\gamma_n)\right]\\ &= R^{(n-1)}\left\{\frac{1}{L}\sum_{\ell=1}^{L} x_\ell(t)\right\} - \frac{1}{L}\sum_{\ell=1}^{L} a_{\ell,n}w(t;\gamma_n)\\ &= R^{(n)}\left\{\frac{1}{L}\sum_{\ell=1}^{L} x_\ell(t)\right\} = R^{(n)}\{y(t)\}. \end{aligned} \tag{7.30}$$

That is, the sum of residual traces, $\sum_\ell R^{(n)}\{x_\ell(t)\}$, is equal to the residual of the trace summation, $R^{(n)}\left\{\sum_\ell x_\ell(t)\right\}$.

This linearity analysis also leads to

$$\begin{aligned}\frac{1}{L}\sum_{\ell=1}^{L}\left\langle w(t;\gamma_n), R^{(n-1)}\{x_\ell(t)\}\right\rangle &= \left\langle w(t;\gamma_n), \frac{1}{L}\sum_{\ell=1}^{L} R^{(n-1)}\{x_\ell(t)\}\right\rangle\\ &= \left\langle w(t;\gamma_n), R^{(n-1)}\left\{\frac{1}{L}\sum_{\ell=1}^{L} x_\ell(t)\right\}\right\rangle\\ &= \left\langle w(t;\gamma_n), R^{(n-1)}\{y(t)\}\right\rangle, \end{aligned} \tag{7.31}$$

where the first equality is derived from the linearity of the product operator $\langle\cdot,\cdot\rangle$, and the second equality is derived from the linearity of the residuum operator. The sum of products across all traces is equal to the product of the wavelet and the sum.

According to Eq. (7.31), the wavelet $w(t;\gamma_n)$ is the best fit to the average of the residual traces. Averaging over multiple traces may effectively suppress data noise and thus stabilise the process. Figure 7.6 demonstrates that the multichannel matching pursuit scheme greatly improves spatial continuity in decomposition and is also able to generate a plausible time–frequency spectrum with high resolution for reservoir detection.

7.6 Structure-adaptive Matching Pursuit

A difference in time delay of a basic wavelet between neighbouring traces will make the average lose its accuracy and will make the extraction of the basic wavelet more complicated. Then, a suboptimal estimation of the basic wavelet will result in more decompositions and lead to inconsistent lateral time–frequency representation. The multichannel matching pursuit can be improved by a structure-adaptive implementation. The best match of the basic wavelet in a local neighbourhood is obtained along the dipping plane, which is defined by the slope of the local structure.

The slope of the local structure can be obtained using the plane-wave destruction method, which assumes the local wavefield to be a plane wave. The plane-wave destructor equation can be expressed in a matrix–vector form as (Wang *et al.*, 2020)

$$\phi(d_x) = \mathbf{e}^{\mathrm{T}}\left[\mathbf{C}(d_x) \circ \mathbf{U}\right]\mathbf{e} = 0, \tag{7.32}$$

where d_x is the apparent slope along the x coordinate, $\mathbf{e}$ is the all-ones vector, $\mathbf{C}(d_x)$ is a two-dimensional operator, $\mathbf{U}$ is the matrix containing data samples, and $\left[\mathbf{C}(d_x) \circ \mathbf{U}\right]$ is the Hadamard product, also known as the element-wise product, of two matrices. The average residual trace in Eq. (7.25) can be rewritten as

$$R^{(n-1)}\{y(t)\} = \frac{1}{L}\sum_{\ell=1}^{L} R^{(n-1)}\{x_\ell(t - d_x x)\}. \tag{7.33}$$

The simultaneous estimation of all wavelet amplitudes of multiple channels is implemented as a multichannel inverse problem with a soft dipping constraint. The objective function is defined in terms of the Frobenius norm as

$$\begin{aligned}\phi(\mathbf{A}_{n\times L}) &= \left\|\mathbf{X}_{N_t\times L} - \mathbf{\Phi}_{N_t\times n}\mathbf{A}_{n\times L}\right\|_F^2 \\ &+\mu\left[\left\|\tilde{\mathbf{X}}^{(n)}_{N_t\times L}\mathbf{D}_x\mathbf{Q}_{\cos}\right\|_F^2 + \left\|\mathbf{Q}_{\sin}\mathbf{D}_t\tilde{\mathbf{X}}^{(n)}_{N_t\times L}\right\|_F^2\right],\end{aligned} \tag{7.34}$$

where $\mathbf{X}_{N_t \times L}$ is the matrix of multichannel seismic data, and $\mathbf{A}_{n \times L}$ is the amplitude matrix, in which each column vector is composed of the wavelet amplitudes of a single trace. The local dipping operators are applied to the reconstruction of multichannel synthetic signals, $\tilde{\mathbf{X}}^{(n)}_{N_t \times L} = \mathbf{\Phi}_{N_t \times n} \mathbf{A}_{n \times L}$, where L is the number of seismic traces. The degree of constraint is controlled softly by the trade-off parameter μ.

In the dipping-structure constraint, $\mathbf{D}_x$ and $\mathbf{D}_t$ are the first order derivative operators along the space axis x and the time axis t, respectively; $\mathbf{Q}_{\cos}$ is a tridiagonal matrix with cosine elements, $\mathbf{Q}_{\cos} = \text{diag}\{\cos\alpha_1, \cos\alpha_2, \cdots, \cos\alpha_{L-1}\}$; $\mathbf{Q}_{\sin}$ is a tridiagonal matrix with the sine elements, $\mathbf{Q}_{\sin} = \text{diag}\{\sin\alpha_1, \sin\alpha_2, \cdots, \sin\alpha_{N_t-1}\}$; and α_i are the dipping angles along the time axis, taken from the central channel of the multiple channels. The Frobenius norm of a matrix is

$$\|\mathbf{R}\|_F^2 = \sum_{i,j} r_{i,j}^2. \tag{7.35}$$

Because it is a sum of all squared elements $r_{i,j}^2$ of matrix $\mathbf{R}$, minimising the objective function $\phi(\mathbf{A}_{n \times L})$ means calculating all wavelet amplitudes in a

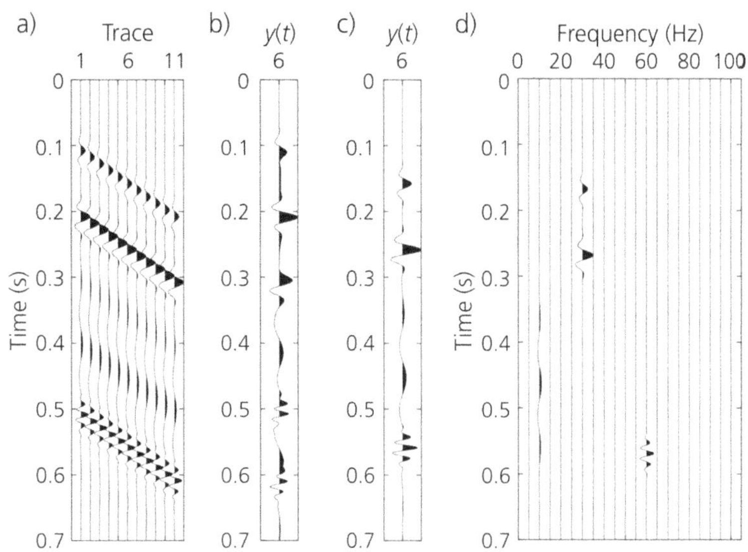

Figure 7.7 Structurally adaptive matching pursuit. (a) A group of synthetic traces. (b) A local average trace computed with a simple average, without structural adaptive constraint. (c) A local average trace computed with a structure-adaptive implementation. (d) Wavelets decomposed by matching pursuit.

least-squares fashion. This procedure not only estimates the amplitude of the current wavelet but also refines the amplitudes of wavelets that were extracted in all previous iterations.

Figure 7.7a displays a group of synthetic traces, and Figures 7.7b and 7.7c are the local average traces that are obtained with a simple average without structure-adaptive constraint and based on dipping information, respectively. It is obvious that the average trace which is obtained using dipping information can better preserve the wavelet information that is shared by all traces.

Figure 7.7d displays the wavelets that are extracted with structure-adaptive implementation, which match the original seismic traces well. The wavelet at around 0.55 s is clearly extracted, which otherwise could be treated as two wavelets at 0.50 s and 0.60 s. This example demonstrates the effectiveness of structure-adaptive implementation in decomposing seismic traces. Because this implementation avoids potential amplitude change and phase distortion caused by the simple averaging, it also accelerates the convergence of decomposition.

7.7 Three Applications

This section shows three application examples of the matching pursuit technique. In the first example, we look at the difference in 'wavelet' between the continuous wavelet transform and the matching pursuit decomposition. Figure 7.8a is an example seismic profile, and Figures 7.8b and 7.8c are two iso-frequency profiles that are generated by these two methods, respectively. In the continuous wavelet transform for a constant dominant frequency f_a, where a is the dilation parameter, the dilated wavelet $\psi_a(t)$ with fixed size is correlated with a seismic trace along the time axis, and the resultant correlation-coefficient gives the continuous spectrum with respect to the time at the given frequency. In matching pursuit, on the contrary, the size of the wavelet varies flexibly along the time axis and is estimated adaptively from the seismic trace. Therefore, matching pursuit shows much higher temporal resolution than that shown by the continuous wavelet transform.

The constituent wavelet in both methods is conceptually different from the wavelet that is used in the convolutional modelling of a seismic trace. The convolutional model assumes a wavelet to be constant within a certain time window, to which a nonstationary filter, accounting for the Earth's

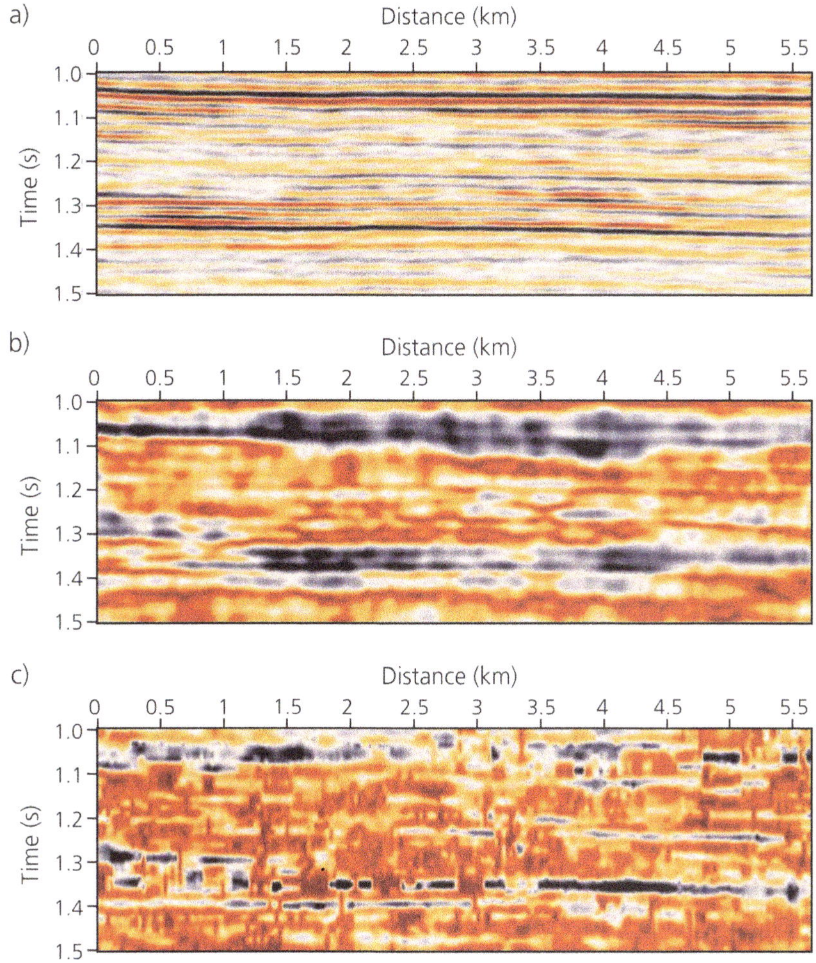

Figure 7.8 Comparison between the continuous wavelet transform and the multichannel matching pursuit method. (a) An example of a seismic profile. (b) The iso-frequency profile (25 Hz) generated by the continuous waveform transform. (c) The iso-frequency profile (25 Hz) generated by the multichannel matching pursuit method.

attenuation effect, can be attached (Eq. 1.7). In the continuous wavelet transform, the wavelet is fixed according to the pseudo-frequency, and thus it is different to the wavelet for the convolutional model.

While the wavelet in matching pursuit is adaptive to the seismic signal, it might be possible to exploit the wavelet concept in the convolutional model and generate a soft constraint along the time axis, additional to the spatial axis, in order to further improve the uniqueness of decomposition by matching pursuit. If this is the case, ideally, the basic wavelets $w(t; \gamma_n)$ that are extracted by matching pursuit could provide the wavelets for the nonstationary convolution model, for inverting the time-dependent reflectivity series.

The second example is to extract a strong coal-seam reflection, so that the remaining target reflections are visible from the seismic profile. As shown in Figure 7.9a, the strong reflection is about 1.5 s across the profile, and the target gas reservoir formation is immediately above it, between 1.4 and 1.5 s.

To track the coherent event, the instantaneous phase information can be used as the input to calculate the cross-correlation between traces. As the instantaneous phase profile of Figure 7.9b shows a clear continuation,

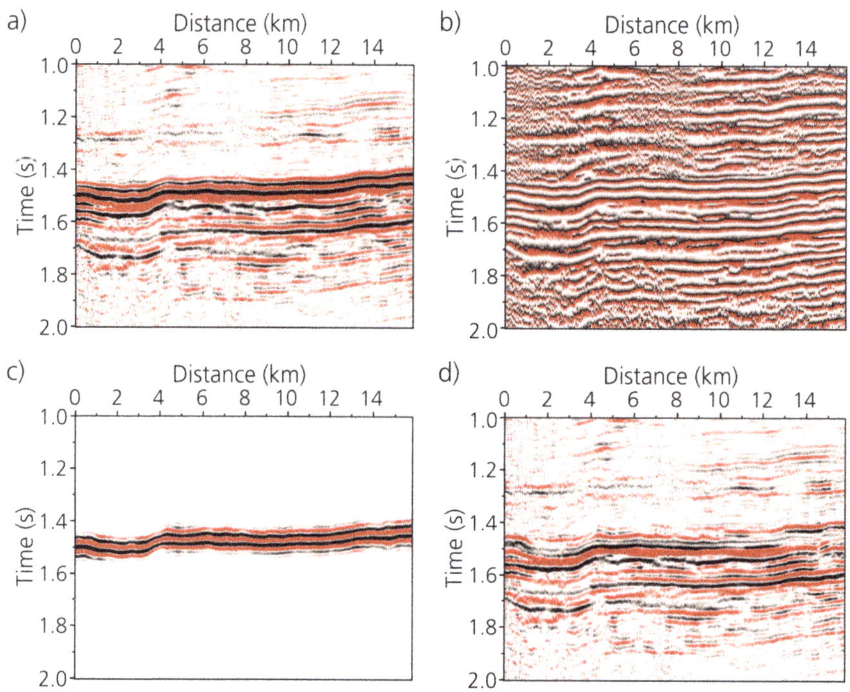

Figure 7.9 Extracting strong coal-seam reflections using multichannel matching pursuit. (a) A seismic profile showing strong coal-seam reflections at about 1.5 s. (b) The instantaneous phase profile that can track the strong coal-seam reflections. (c) The strong coal-seam reflections extracted by multichannel matching pursuit. (d) The seismic profile after removal of the coal-seam reflections, revealing the hidden weak reflections.

following the horizon at a time of about 1.5 s, performing cross-correlation between phase traces can automatically track the horizon. This correlation provides us with the initial time abscissa u_ℓ, where ℓ is the trace index.

Figure 7.9c displays the coal-seam reflections that are extracted along the horizon, by the multichannel matching pursuit. In this application, the number of traces within a spatial window is set to be either 3 or 5. Both settings produce similar results.

The profile in Figure 7.9d is the same as that in Figure 7.9a, but after the removal of the strong coal-seam reflections. That is, it is the difference between Figures 7.9a and 7.9c. This profile clearly shows weak reflections across it at the time interval between 1.4 and 1.5 s. These target reflections were hidden by strong coal-seam reflections. The multichannel matching

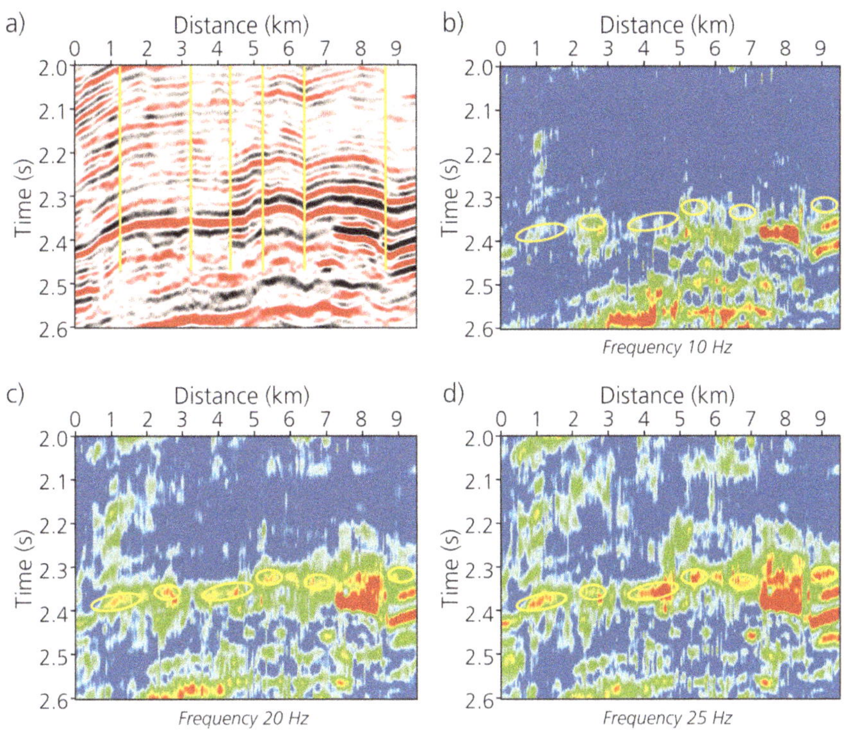

Figure 7.10 Detecting low-frequency shadows of gas reservoirs by multichannel matching pursuit. (a) A crooked seismic line across six wells, represented by vertical lines. (b–d) Three iso-frequency profiles at 10, 20, and 25 Hz. Below the target gas reservoirs (the ellipses), the low-frequency shadows are gradually reduced at 10 Hz, while the target gas reservoirs have strong amplitudes at 20 and 25 Hz.

pursuit algorithm exploits the spatial continuity to remove the strong coal-seam reflections from the seismic profile, to allow the weak target reflections, that are immediately on the top of coal-seam reflections, to be characterised.

The third example is to generate a time–frequency spectrum to detect gas reservoirs. Low-frequency shadows in the time–frequency spectrum are direct indicators for gas reservoir detection (Chakraborty & Okaya, 1995; Castagna *et al.*, 2003; Korneev *et al.*, 2004).

Figure 7.10a shows a crooked seismic line across six wells, and Figures 7.10b–d display three frequency profiles, that are obtained by the multichannel matching pursuit algorithm. We can see that, below the target gas reservoirs (yellow circles), the low-frequency shadows at 10 Hz are gradually reduced, but, meanwhile, the target gas reservoirs show strong amplitudes at 20 and 25 Hz.

In summary, the multichannel matching pursuit algorithm greatly improves spatial continuity in seismic signal decomposition and the temporal resolution of the resultant time–frequency spectrum. Multichannel matching pursuit is also able to generate a plausible time–frequency spectrum for detecting low-frequency shadows underneath gas reservoirs. In addition, the matching pursuit-based time–frequency spectrum, unlike the standard WVD, does not suffer from cross-term interference, because the WVD of each individual wavelet that is extracted by matching pursuit is calculated separately.

8

Local Power Spectra with Multiple Windows

In the time–frequency representation, the frequency spectrum at a particular time position is a local property. But it is not exactly an instantaneous property. This is because the spectral analysis must be performed over a segment of the signal, and not over an isolated sample of the signal. A segment of the signal always has a certain duration of oscillation, no matter how short the duration is.

Time–frequency analysis methods often involve a time window function that serves as a weighting function by which the signal is multiplied by to form a segment. For example, the Gabor transform-type methods, which evolved from the short-time Fourier transform, make various modifications to the window function. The continuous wavelet transform generalises the complex Gabor wavelet defined with a Gaussian function. The Wigner–Ville distribution (WVD) also applies a time window function to the signal, to smooth out the cross-term interference between the different signal components from the WVD spectrum.

The central sample of the signal segment is often assigned the highest weighting coefficient. The rest of the segment is not adequately represented in the frequency spectrum. Therefore, the time–frequency spectrum loses information about the signal segment because of the application of the time window function.

While we acknowledge that the time–frequency spectrum is a local property and is not an instantaneous property at a particular time position, we also admit that the time–frequency spectrum does not properly

Time–Frequency Analysis of Seismic Signals, First Edition. Yanghua Wang.

represent the local properties of the nonstationary seismic signal. Therefore, in this chapter, we propose a method to properly represent the local spectral properties in the time–frequency plane, by using multiple orthogonal window functions, and recover the information lost by time windowing before applying the Fourier transform.

8.1 Multiple Orthogonal Windows

The information lost in time–frequency analysis can be recovered by using a scheme with multiple orthogonal windows. In the spectral theory of the conventional one-dimensional Fourier transform, this scheme with multiple windows is called the multitaper method (Thomson, 1982).

Consider a complex harmonic component

$$x(t) = \hat{x}(f_0)e^{i2\pi f_0 t}, \tag{8.1}$$

where f_0 is a constant frequency. Multiplying $x(t)$ by a window function $w(t)$ forms a new signal:

$$y(t) = w(t)x(t). \tag{8.2}$$

The Fourier transform of the windowed harmonic component $y(t)$ is

$$\hat{y}(f) = \int_{-T}^{T} w(t)\hat{x}(f_0)e^{-i2\pi(f-f_0)t}dt, \tag{8.3}$$

where T is the (half-)length of the time window function $w(t)$.

The window function $w(t)$ is not an arbitrary function, but rather it is an optimal function which maximises the energy within the frequency band between $[f_0 - B]$ and $[f_0 + B]$, where B is half of the frequency bandwidth and, therefore, defines the spectral resolution. If the energy within the frequency band is normalised by the energy in the entire frequency range, we define the bandwidth retention factor λ as follows:

$$\lambda = \frac{\int_{f_0-B}^{f_0+B} |\hat{y}(f)|^2 df}{\int_{-\infty}^{\infty} |\hat{y}(f)|^2 df}. \tag{8.4}$$

This factor is always less than unity (Park *et al.*, 1987). An optimal window function $w(t)$ should have the factor λ close to unity.

The numerator of the ratio factor can be evaluated as

$$\begin{aligned}
\int_{f_0-B}^{f_0+B} |\hat{y}(f)|^2 \mathrm{d}f &= \int_{f_0-B}^{f_0+B} \left(\int_{-T}^{T} w(t)\hat{x}(f_0)\mathrm{e}^{-\mathrm{i}2\pi(f-f_0)t}\mathrm{d}t \right)^* \left(\int_{-T}^{T} w(\tau)\hat{x}(f_0)\mathrm{e}^{-\mathrm{i}2\pi(f-f_0)\tau}\mathrm{d}\tau \right) \mathrm{d}f \\
&= |\hat{x}(f_0)|^2 \int_{-T}^{T} w^*(t)\mathrm{d}t \int_{-T}^{T} w(\tau)\mathrm{d}\tau \int_{f_0-B}^{f_0+B} \mathrm{e}^{\mathrm{i}2\pi(f-f_0)(t-\tau)}\mathrm{d}f \\
&= |\hat{x}(f_0)|^2 \int_{-T}^{T} w^*(t)\mathrm{d}t \int_{-T}^{T} w(\tau)\mathrm{d}\tau \int_{-B}^{B} \mathrm{e}^{\mathrm{i}2\pi f(t-\tau)}\mathrm{d}f \\
&= |\hat{x}(f_0)|^2 \int_{-T}^{T} w^*(t)\mathrm{d}t \int_{-T}^{T} b(t-\tau)w(\tau)\mathrm{d}\tau,
\end{aligned} \tag{8.5}$$

where $b(t-\tau)$ is a scaled sinc function:

$$b(t-\tau) = \int_{-B}^{B} \mathrm{e}^{\mathrm{i}2\pi f(t-\tau)}\mathrm{d}f = 2B\left(\frac{\sin[2\pi B(t-\tau)]}{2\pi B(t-\tau)} \right). \tag{8.6}$$

The denominator of the ratio factor can be evaluated following Parseval's theorem:

$$\begin{aligned}
\int_{-\infty}^{\infty} |\hat{y}(f)|^2 \mathrm{d}f = \int_{-\infty}^{\infty} |y(t)|^2 \mathrm{d}t &= |x(t)|^2 \int_{-T}^{T} w^*(t)w(t)\mathrm{d}t \\
&= |\hat{x}(f_0)|^2 \int_{-T}^{T} w^*(t)w(t)\mathrm{d}t.
\end{aligned} \tag{8.7}$$

Then, the bandwidth retention factor λ can be expressed as follows:

$$\lambda = \frac{\int_{-T}^{T} w^*(t)\left[\int_{-T}^{T} b(t-\tau)w(\tau)\mathrm{d}\tau \right]\mathrm{d}t}{\int_{-T}^{T} w^*(t)w(t)\mathrm{d}t}. \tag{8.8}$$

This is an optimisation problem, in which maximising λ is equivalent to minimising $(1-\lambda)$, a quantity of spectral leakage.

Equation (8.8) can be simplified to the Fredholm integral equation (Slepian, 1954, 1964, 1978) as follows:

$$\int_{-T}^{T} b(t-\tau)w(\tau)\mathrm{d}\tau = \lambda w(t). \tag{8.9}$$

The Fredholm integral equation recasts the optimisation problem of Eq. (8.8) as an eigenfunction problem, in which the eigenvalue is the bandwidth retention factor λ, and the corresponding eigenfunction is the time window function $w(t)$.

8.2 Multiple Windows Defined by the Prolate Spheroidal Wavefunctions

The solutions $w(t)$ of Eq. (8.9) are scaled versions of the prolate spheroidal wavefunctions (Slepian & Pollak, 1961; Slepian, 1964). The prolate spheroidal wavefunctions (Niven, 1880; Nicholson, 1925; Van Buren & Boisvert, 2002, 2004) are the solutions of the following differential equation:

$$\frac{\mathrm{d}}{\mathrm{d}z}\left((1-z^2)\frac{\mathrm{d}}{\mathrm{d}z}U_{m,n}(c,z)\right)+\left(\chi_{m,n}(c)-c^2z^2-\frac{m^2}{1-z^2}\right)U_{m,n}(c,z)=0, \tag{8.10}$$

where z is a scaled variable within the closed interval $z\in[-1,1]$, c is the spheroidal wave parameter, $\chi_{m,n}(c)$ are the eigenvalues, $U_{m,n}(c,z)$ are the prolate spheroidal wavefunctions, n and m are two integer indexes, and $0\le m\le n$. In this differential equation, both $\chi_{m,n}(c)$ and $U_{m,n}(c,z)$ depend on the spheroidal wave parameter c, which is the product of the time limit T and the frequency band B (Slepian, 1964).

Equation (8.10) is a scalar Helmholtz equation and is transferred from the rectangular coordinates to the general ellipsoidal coordinates (Flammer, 1957). The solutions of Eq. (8.10), which are the prolate spheroidal wavefunctions $U_{m,n}(c,z)$, are summarised in Appendix F. But note here that the eigenvalues $\chi_{m,n}(c)$ are not the same as the eigenvalues λ defined in the Fredholm integral equation (Eq. 8.9).

The prolate spheroidal wavefunctions $U_{m,n}(c,z)$, which are orthogonal to each other, define the window functions we are looking for:

$$w_{p=n-m}\left(c,\frac{t}{T}\right)=\frac{1}{r_{m,n}(c)}U_{m,n}\left(c,\frac{t}{T}\right), \tag{8.11}$$

where $r_{m,n}(c)$ is a normalisation factor,

$$r_{m,n}(c) = \left(\int_{-T}^{T} U_{m,n}^2\left(c, \frac{t}{T}\right) \mathrm{d}t \right)^{1/2}. \tag{8.12}$$

The prolate spheroidal wavefunctions $U_{m,n}(c, z)$ are the associated Legendre polynomials (Nicholson, 1925), which in general are not orthogonal to each other. For example, $U_{1,1}(c, z)$ is not orthogonal to $U_{2,2}(c, z)$. However, they satisfy the orthogonality condition either for a fixed m or for a fixed n, assuming that $0 \le m \le n$.

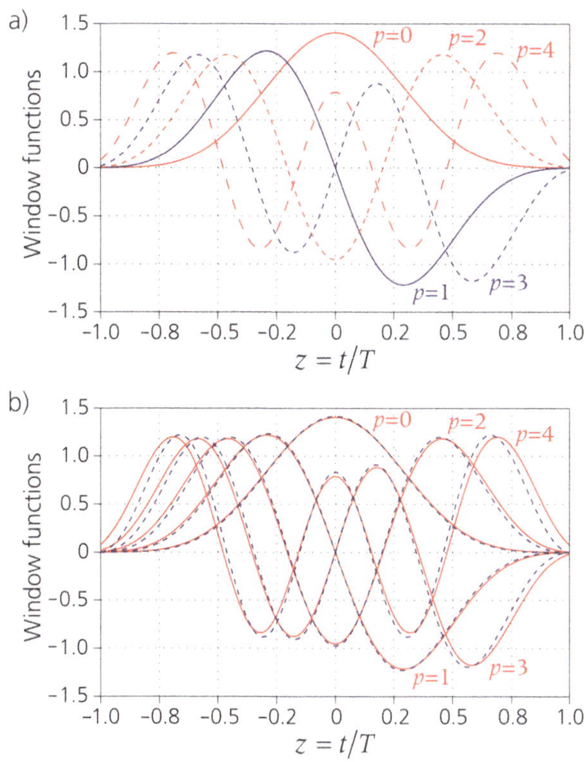

Figure 8.1 Multiple windows defined by the prolate spheroidal wavefunctions. (a) Five orthogonal window functions $w_p(c, z)$, that are scaled versions of the prolate spheroidal wavefunctions $U_{m,n}(c, z)$, where $m = 0$ and $p = n - m$. The red and blue curves correspond to even numbers of zero crossings, $p = \{0, 2, 4\}$, and odd numbers of zero crossings, $p = \{1, 3\}$, respectively. (b) Comparison of the window functions $w_p(c, z)$ formed by the prolate spheroidal wavefunctions $U_{m,n}(c, z)$ with $m = 0$ (solid curves) and $m = 3$ (dashed curves).

When these orthogonal functions form an orthogonal basis, the seismic signal can be expressed as a linear combination of these independent window functions. Figure 8.1a displays five window functions $w_p(c,z)$, where $z = t/T$, $m = 0$, and $p = n - m = \{0, 1, \cdots, 4\}$. The red curves correspond to even orders of $p = \{0, 2, 4\}$, and the blue curves correspond to odd orders of $p = \{1, 3\}$. Figure 8.1a indicates that the order p is the number of zero crossings within the interval $z \in (-1, 1)$.

Figure 8.1b reveals that the window functions $w_p(c,z)$ formed by the prolate spheroidal wavefunctions with different m values are similar. The red solid curves are the window functions built with $m = 0$, and the blue dashed curves are the window functions built with $m = 3$.

Figure 8.1 also suggests that window $w_p(c,z)$ is an even function if $p = n - m$ is an even order, and window $w_p(c,z)$ is an odd function if $p = n - m$ is an odd order.

The spectral estimate based on the window $w_{p=0}(c,z)$ depends mainly on the central part of the signal segment, as is the case with conventional window functions. But the spectral estimates based on higher order window functions $w_{p\geq 1}(c,z)$ depend more on the energy at the beginning and end of the signal segment. Therefore, the combination of the spectral estimates that are generated with several window functions can largely recover the local spectrum of the signal.

8.3 Multiple Windows Constructed by Solving a Discretised Eigenvalue Problem

Alternatively, the Fredholm integral equation (Eq. 8.9) can be discretised and represented in a matrix–vector form as

$$\mathbf{B}\mathbf{w} = \lambda \mathbf{w}, \tag{8.13}$$

where $\mathbf{B}$ is the matrix with elements defined by the scaled sinc function $b(t - \tau)$, and $\mathbf{w}$ is the vector of a discretised window. The Fredholm integral equation becomes a discretised eigenvalue problem, where the eigenvalue λ is the bandwidth retention factor, and the corresponding eigenvector $\mathbf{w}$ is the discretised window.

Since $\mathbf{B}$ is a positive-definite Toeplitz matrix, all the eigenvalues are real and positive, and they can be listed in descending order:

$$\lambda_0 \geq \lambda_1 \geq \lambda_2 \geq \cdots. \tag{8.14}$$

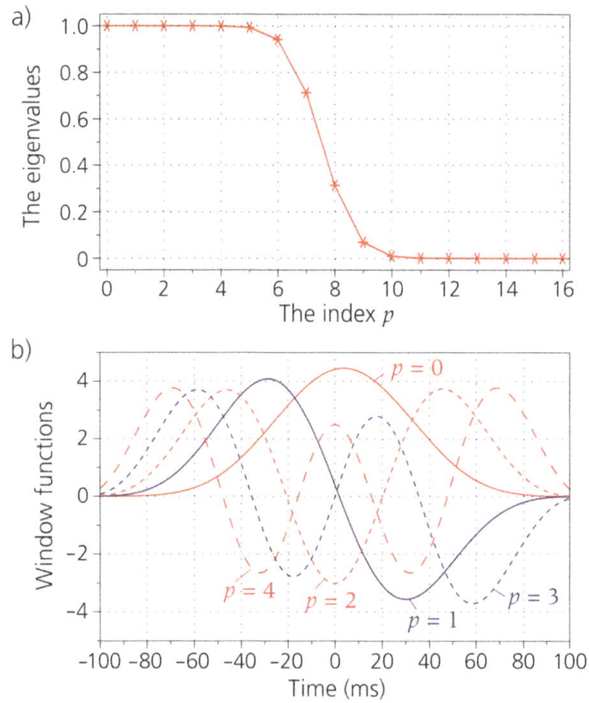

Figure 8.2 Multiple windows constructed by solving a discretised eigenvalue problem. (a) The eigenvalues of the Toeplitz matrix **B**. The eigenvalues λ_p, with index p, are shown in descending order. (b) Five optimal window functions $w_p(t)$, for $p = \{0, 1, \cdots, 4\}$, formed by the five eigenvectors corresponding to the five highest eigenvalues.

The larger eigenvalues indicate that the corresponding eigenvectors $\mathbf{w}_p \equiv w_p(t)$ are exactly the optimal window functions we are looking for. All window functions are band-limited, and the (half-)bandwidth is B.

In constructing the Toeplitz matrix **B**, we use a scaled sinc function $b(t-\tau)$ truncated at $|2\pi B(t-\tau)| = 4\pi$, which corresponds to $c = \pi BT \le 2\pi$ or $BT \le 2$. Solving the discretised eigenvalue problem of Eq. (8.13), we obtain the maximum λ values, and seek the corresponding eigenvectors $\mathbf{w}$ which attain these maximal values.

Figure 8.2a shows the eigenvalues obtained. The five largest eigenvalues are

$$\lambda_p = \{1.0,\ 1.0,\ 0.999999,\ 0.999971,\ 0.999465\}. \tag{8.15}$$

Figure 8.2b shows the corresponding window functions $w_p(t)$. These five window functions are mutually orthogonal,

$$\phi_{m,n} = \int_{-T}^{T} w_m(t) w_n(t) \mathrm{d}t \approx 0, \qquad \text{for } m \neq n, \tag{8.16}$$

because they are the eigenvectors of the Toeplitz matrix **B**.

The window functions $\mathbf{w}_0 = w_0(t)$ and $\mathbf{w}_1 = w_1(t)$, corresponding to the maximum eigenvalues λ_0 and λ_1, have almost no spectral leakage. The window functions $\mathbf{w}_p = w_p(t)$, for $p = \{2, 3, 4\}$, corresponding to the eigenvalues $\lambda_p \approx 1$, have only small spectral leakage. The energy of a window function was scaled to unity.

The zeroth order window function $\mathbf{w}_0 = w_0(t)$ resembles, but is not identical to, the truncated Gaussian window. It is almost symmetrical around the central peak, and tapers to zero at the ends (Wilson, 1987; Wilson & Spann, 1988).

Higher order window functions are symmetric: those with even p orders are symmetric about the centre, while those with odd p orders are antisymmetric. The window function with $p = 1$ contains the energy concentrated in the early part of the signal segment, and the window function with a higher p contains the energy concentrated even earlier.

Figure 8.3 compares the window functions defined by the prolate spheroidal wavefunctions (solid red curves) with those constructed by the eigenvectors of the Toeplitz matrix **B** (dashed blue curves). This direct

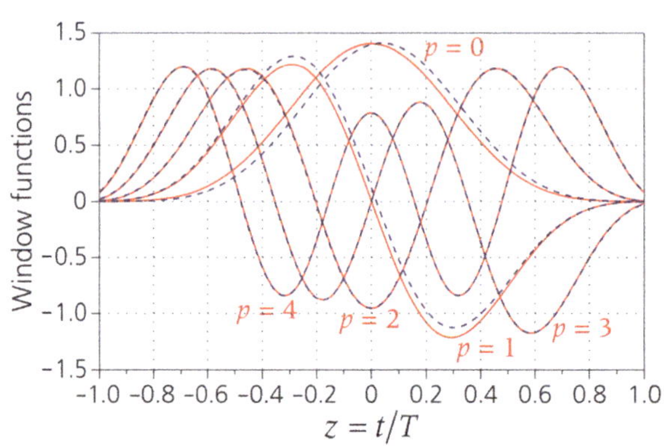

Figure 8.3 Comparison between the window functions defined in terms of the prolate spheroidal wavefunctions (solid red curves) and those constructed by the eigenvectors of the Toeplitz matrix **B** (dashed blue curves).

comparison shows that, for low orders $p = \{0, 1\}$, the window functions $w_p(t)$ that are constructed by the eigenvectors of the Toeplitz matrix $\mathbf{B}$ are skewed, while the window functions that are defined in terms of the prolate spheroidal wavefunctions are either symmetric or antisymmetric.

To allow direct comparison, we have imposed conventions for a scaling and a polarity convention in Figure 8.3. For the scaling, as stated in previous chapters, the energy of each window function is set to unity:

$$\int_{-T}^{T} \left|w_p(t)\right|^2 \mathrm{d}t = 1. \tag{8.17}$$

For an even-order window function, the average is made positive:

$$\int_{-T}^{T} w_p(t)\mathrm{d}t > 0. \tag{8.18}$$

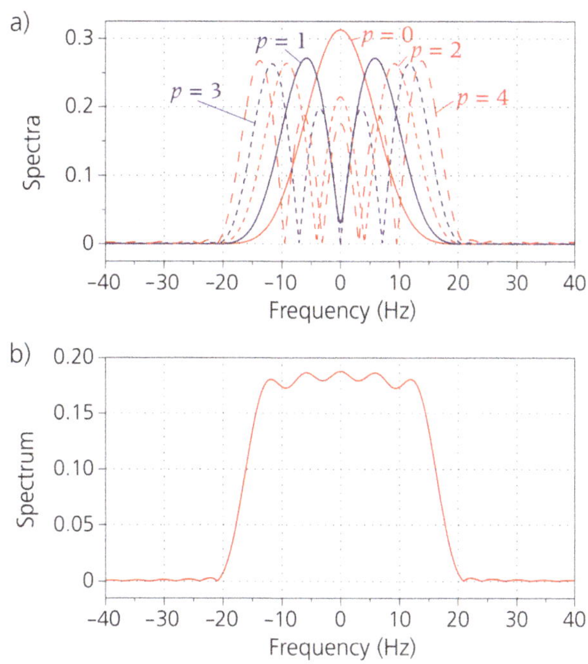

Figure 8.4 The amplitude spectra of multiple windows constructed by the discretised eigenvalue problem. (a) The amplitude spectra of five optimal window functions. The order of the eigenvalues is denoted as p. (b) The smoothed amplitude spectrum obtained by a weighted average of the five power spectra.

For an odd-order window function, the first lobe is made positive:

$$\int_{-T}^{T} (T - t) w_p(t) \mathrm{d}t > 0. \tag{8.19}$$

Figure 8.4 shows the amplitude spectra of the window functions $\left|\hat{w}_p(f)\right|$, $p = \{0, 1, \cdots, 4\}$. The smoothed amplitude (square-root of the power) spectrum is obtained by a weighted average of these power spectra:

$$\left|\hat{w}(f)\right| = \left(\frac{\sum_{p=0}^{P} \lambda_p \left|\hat{w}_p(f)\right|^2}{\sum_{p=0}^{P} \lambda_p} \right)^{1/2}, \tag{8.20}$$

where $P = 4$ in this example, and each eigenvalue λ_p is approximately equal to 1. The smoothed spectrum has a maximum flat spectrum within the frequency band $[-B, B]$.

8.4 Multiple Windows Constructed by Gaussian Functions

Theoretically, both methods, either the prolate spheroidal wavefunctions, which are the solutions of a Helmholtz equation, or the solution of the discretised eigenvalue problem, can yield a complete set of orthogonal window functions. However, if we use only five window functions, corresponding to the five largest λ values, the passband of the average spectrum is not flat enough, as can be seen in Figure 8.4b. We need to include more window functions in the weighted average, to improve the spectral flatness within the frequency passband.

Furthermore, solving the Helmholtz equation for the prolate spheroidal wavefunctions, or solving the discretised eigenvalue problem for the eigenvectors, is not the efficient method we need in practice, as we often need on-the-fly calculation of the windows in signal analysis. Therefore, instead of the sophisticated numerical calculations presented in the previous sections, I propose to construct multiple window functions directly.

I construct the multiple window functions using a stack of Gaussian functions:

$$w_p(t) = \frac{1}{r_p} \sum_{m=0}^{p} (-1)^m \mathrm{e}^{-\frac{1}{2\sigma^2 T^2}\left((2+p)t - \frac{2(2m-p)}{B}\right)^2}, \tag{8.21}$$

where p is the order of the window function, T is the (half-)window length, B is half of the frequency bandwidth, σ is the standard deviation of a Gaussian function, and r_p is the normalisation factor:

$$r_p = \left[\frac{\sqrt{\pi}\sigma T}{2+p}\left(1+p+2\sum_{m=0}^{p-1}\sum_{n=m+1}^{p}(-1)^{m+n}e^{-\frac{4(m-n)^2}{\sigma^2 B^2 T^2}}\right)\right]^{1/2}. \tag{8.22}$$

The normalisation factor makes the energy of the window function $w_p(t)$ unity (Eq. 8.17). Therefore, the energy of the signal segment is not affected when the segment is multiplied by the window function.

Here, I set the maximum order of the window function to $P = 4$. For the case of $BT = 2$, the standard deviation is set as $\sigma = 0.72$, and the window

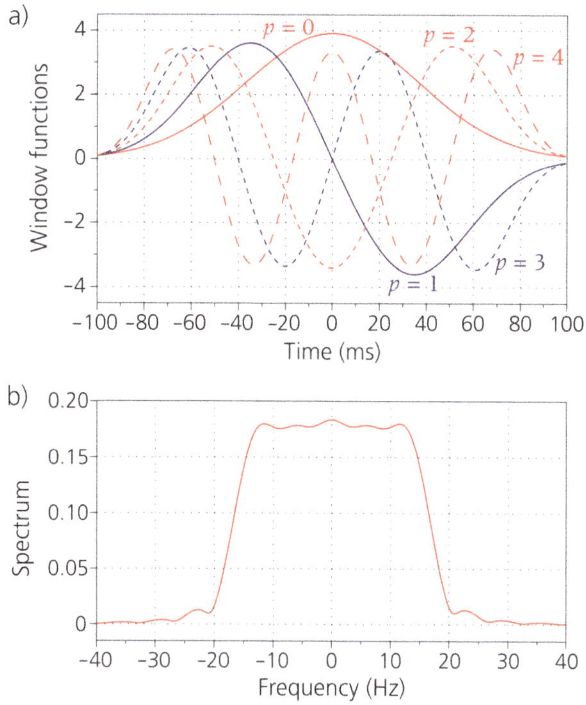

Figure 8.5 Multiple windows constructed by Gaussian functions, with the (half-) bandwidth $B = 20$ Hz and $(BT, \sigma) = (2, 0.72)$. (a) Five windows formed by a group of Gaussian functions, $p = \{0, 1, \cdots, 4\}$. (b) The weighted average spectrum of these five window functions. Averaging the spectra produced by these five windows gives an excellent flat spectrum.

functions $w_p(t)$ that are defined within the time interval $[-T, T]$ produce a spectrum with a flat passband. Figure 8.5a displays five windows formed by a stack of Gaussian functions, with the (half-)bandwidth $B = 20$ Hz, and $(BT, \sigma) = (2, 0.72)$. The values of the bandwidth retention factor of these five window functions are close to the desirable maximum value $\lambda = 1$:

$$\lambda = \{0.999029, 0.999446, 0.999622, 1.000070, 1.000200\}. \tag{8.23}$$

This shows that the factor λ does not necessarily decrease as the order of the window function increases. Therefore, the assumption contained in Equation (8.14), that the window functions were formed by solving the eigenvalue problem, no longer applies.

Figure 8.5b shows the weighted average spectrum of these five window functions. The figure clearly demonstrates that this set of orthogonal windows can produce an average spectrum with a flat passband, as opposed to the spectrum shown in Figure 8.4b, the weighted average spectrum of five eigenvectors.

However, these windows formed by a stack of Gaussian functions are not strictly orthogonal to each other. The cross-correlations between any pair of window functions, calculated by Eq. (8.16), are listed as follows:

$$\begin{bmatrix} \phi_{0,1} & \phi_{0,2} & \phi_{0,3} & \phi_{0,4} \\ & \phi_{1,2} & \phi_{1,3} & \phi_{1,4} \\ & & \phi_{2,3} & \phi_{2,4} \\ & & & \phi_{3,4} \end{bmatrix} = \begin{bmatrix} 0 & -0.043 & 0 & 0.027 \\ & 0 & -0.080 & 0 \\ & & 0 & -0.085 \\ & & & 0 \end{bmatrix}. \tag{8.24}$$

The cross-correlations $\phi_{m,n}$ indicate that these window functions are not completely orthogonal but are approximations.

Figure 8.6 shows another set of window functions formed by a stack of Gaussian functions, with the (half-)bandwidth of $B = 40$ Hz and $(BT, \sigma) = (2, 0.72)$. The weighted average spectrum of these five window functions also has a flat central passband, although it does have small sidelobes at both edges of the passband.

While the aggregate sum of several windows gives a flat passband, the energy level of the sidelobes of the spectrum is also an important property of a window function. Large sidelobes can potentially mask some of the weak events (Bekara, 2013). Theoretically, for a Gaussian function with infinite duration, there are no sidelobes in the frequency spectrum. Due to

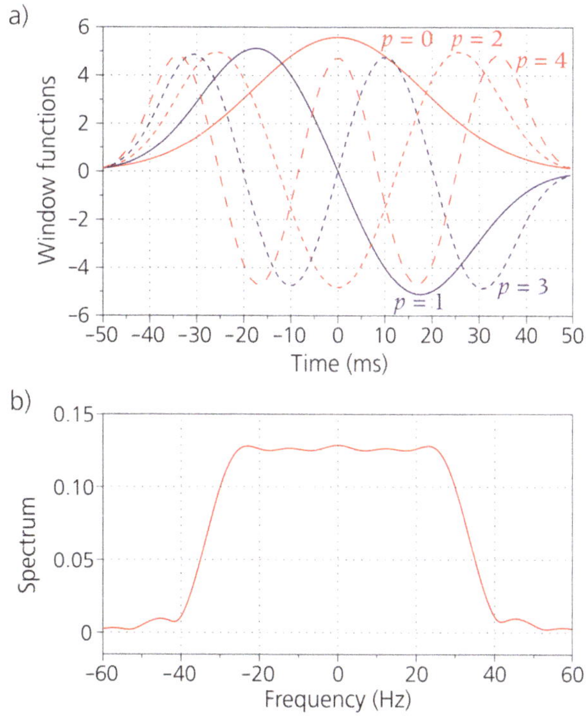

Figure 8.6 Multiple windows constructed by Gaussian functions, with the (half-) bandwidth $B = 40$ Hz and $(BT, \sigma) = (2, 0.72)$. (a) Five windows constructed by a group of Gaussian functions. (b) The weighted average spectrum of the five window functions.

the finite duration of either the window function or the signal segment itself, the truncation effect leads to sidelobes in the spectrum. However, the largest sidelobe of a truncated Gaussian function is much smaller than for a rectangular window with or without tapering (Chapter 2).

8.5 The Gabor Transform with Multiple Windows

The Gabor transform-type method can be considered as a cross-correlation between the signal and various complex wavelets. The complex wavelet, also called the Gabor wavelet, in the standard Gabor transform is the product of the kernel of the Fourier transform and a time window function. The window function in forming the Gabor wavelet, which is defined as a

Gaussian function, is translated across the signal. In the same way, a dilated complex wavelet in the continuous wavelet transform is also translated across the signal. Due to the application of the window function in the formation of the complex wavelet, there is a loss of information in the frequency spectrum at any given time position. If we use the multiple orthogonal windows, we can retain the spectral information within the frequency band.

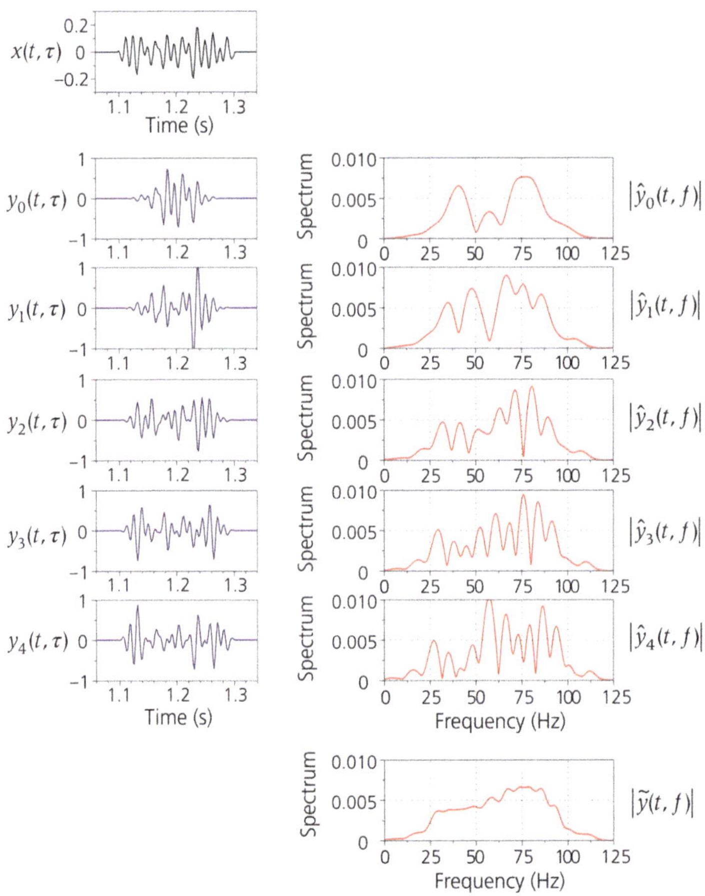

Figure 8.7 Spectral analysis with multiple windows, at a given time t. The original signal segment is $x(t,\tau)$ (the black signal at the top), with the reference time position at $t = 1.2$ s and a (half-)length of $T = 0.1$s. The segments formed by the five window functions are $y_p(t,\tau)$ (the blue signals). The average of five time–frequency spectra (in red) gives a smooth amplitude spectrum $|\tilde{y}(t,\tau)|$.

Figure 8.7 illustrates the implementation of the Gabor transform with multiple orthogonal windows, at a given time position t. The original signal segment (the black curve in the upper-left corner) is $x(t,\tau)$ at time $t = 1.2$ s. The (half-)length of the duration of time τ is $T = 0.1$ s. A weighted segment (the blue curve) with any one of five orthogonal window functions is $y_p(t,\tau)$, where $p = \{0, 1, \cdots, P\}$, and $P = 4$ is the highest order selected.

Recall that the bandwidth retention factor λ is quantified in terms of the power spectrum (Eq. 8.4). Therefore, different λ values should be applied as weighting coefficients to the corresponding power spectra $\left|\hat{y}_p(t,f)\right|^2$. The final amplitude spectrum $\left|\tilde{y}(t,f)\right|$ is a weighted average of these power spectra:

$$\left|\tilde{y}(t,f)\right| = \left(\frac{\sum_{p=0}^{P} \lambda_p \left|\hat{y}_p(t,f)\right|^2}{\sum_{p=0}^{P} \lambda_p} \right)^{1/2}. \tag{8.25}$$

The final amplitude spectrum $\left|\tilde{y}(t,f)\right|$ shown in Figure 8.7 is a smooth spectrum with less spectral leakage than $\left|\hat{y}_0(t,f)\right|$, the spectrum of a conventional single window scheme.

Figure 8.8 compares the time–frequency spectra produced by the Gabor transform when either a single window function or multiple window functions are used. In the calculation, the same length $T = 50$ ms is used for each window function. If the Gabor transform uses just a single window function (Figure 8.8b), the window does not properly represent the contribution of the signal segment towards to both ends of the time duration, and the time–frequency spectrum does not represent all the information within the signal segment. When the Gabor transform uses multiple windows (Figure 8.8c), the information that is lost when implemented with only the single window scheme ($p = 0$) is restored by the other window functions ($p \geq 1$).

The Gabor transform is a typical linear transformation method. One of the advantages of the linear transformation is that the time–frequency spectrum can be used to reconstruct the original signal. However, with the multi-window Gabor transform, the resultant time–frequency spectrum is a local amplitude spectrum, $\left|\tilde{y}(t,f)\right|$, based on which the original signal cannot be reconstructed. Instead, the smooth time–frequency spectrum shows a clear picture of the frequency band, varying along the time axis. Based on the time–frequency spectrum, we can statistically evaluate the mean frequency and the standard deviation. The mean frequency function $f_m(t)$ is evaluated from the time–frequency spectrum by

$$f_m(t) = \frac{1}{E(t)} \int_0^{\infty} \mathrm{d}f \int_{-T}^{T} f \left| \tilde{y}(t-\tau, f) \right|^2 L(\tau)\, \mathrm{d}\tau , \tag{8.26}$$

and the standard deviation function $f_\sigma(t)$ is evaluated by

$$f_\sigma(t) = \left(\frac{1}{E(t)} \int_0^{\infty} \mathrm{d}f \int_{-T}^{T} (f - f_m)^2 \left| \tilde{y}(t-\tau, f) \right|^2 L(\tau) \mathrm{d}\tau \right)^{1/2} , \tag{8.27}$$

where $\left| \tilde{y}(t, f) \right|^2$ is the power spectrum at time t, $E(t)$ is the instantaneous energy,

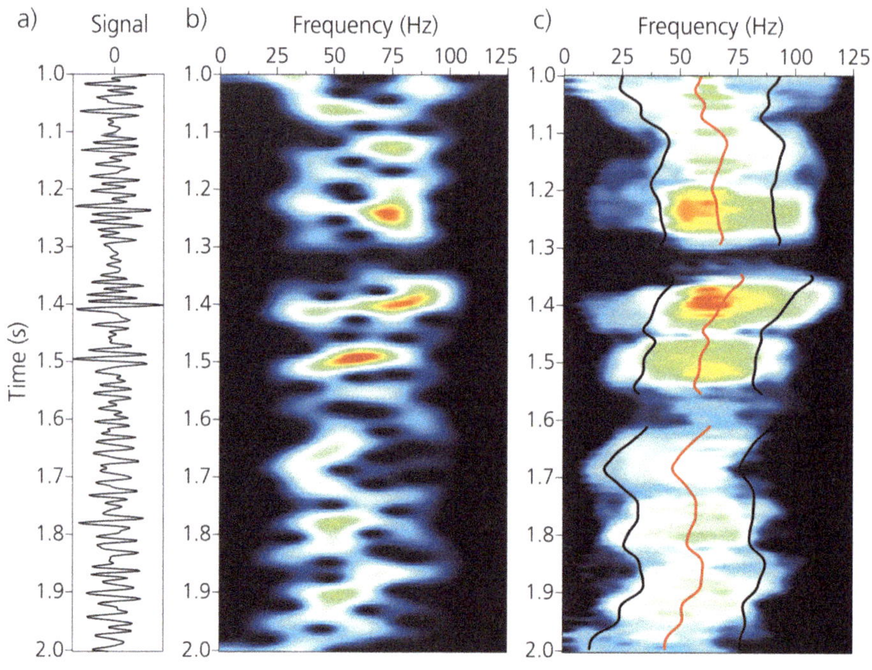

Figure 8.8 The Gabor transform using a single or multiple window function. (a) An example of a field seismic trace. (b) The time–frequency spectrum produced by the Gabor transform using the Blackman–Harris window. (c) Local power spectrum generated by the Gabor transform with multiple orthogonal windows. The red curve is the mean frequency $f_m(t)$, and the two black curves are $f_m(t) \pm f_\sigma(t)$, where $f_\sigma(t)$ is the standard deviation.

$$E(t) = \int_{0}^{\infty} \mathrm{d}f \int_{-T}^{T} \left|\tilde{y}(t-\tau, f)\right|^2 L(\tau)\mathrm{d}\tau, \tag{8.28}$$

and $L(t)$ is a low-pass filter that is applied to the time axis. For the statistical evaluations in Eqs. (8.26) and (8.27), a pragmatic condition is set as $E(t) \geq E_\ell$, where E_ℓ is a threshold of the energy.

Figure 8.8c also shows a stable estimate for the mean frequency and the standard deviation from the time–frequency spectrum. The red curve is the mean frequency function $f_m(t)$, and the two black curves are $f_m(t) \pm f_\sigma(t)$, respectively. According to the definition of Eq. (8.4), the bandwidth retention factor λ is maximised at any time position within the band $[f_0 - B, f_0 + B]$. Therefore, the evaluated standard deviation function $f_\sigma(t)$ has only a small variation with respect to time t.

8.6 The WVD with Multiple Windows

The WVD method analyses the instantaneous autocorrelation of the analytic signal and generates the energy density distribution in the time–frequency plane. The standard WVD method suffers from the cross-term interference between the signal components. Since this cross-term interference has both positive and negative amplitudes, it can be suppressed by two-dimensional smoothing in the time–frequency plane.

In the two-dimensional smoothing process, the frequency-smoothing filter $\hat{h}(f)$ can be represented in the time domain as $h(\tau)$, which is equivalent to the WVD of a window function $k(t)$. This window function can be applied directly to the analytic signal $z(t)$, before the instantaneous autocorrelation $\tilde{\phi}(t, \tau)$ is calculated, prior to the Fourier transform is performed with respect to time τ.

When the analytic signal $z(t)$ is multiplied by the window function $k(t)$, the latter has the highest weight in the centre of the window, when the window function is defined with a Gaussian function. Due to the tapering effect on the analytic signal $z(t)$, there is a loss of information in the WVD spectrum produced by the local Fourier transform of the windowed signal, $\tilde{z}(t) = k(t)z(t)$ (Eq. 6.46). Therefore, I suggest using multiple orthogonal windows, rather than only a single window, in the WVD generation.

At any given time t, taking advantage of the linearity of the Fourier transform, we can apply each window directly to the analytic signal $z(t)$, to produce the instantaneous autocorrelation accordingly, and then form an averaged instantaneous autocorrelation by a weighted sum, before performing the Fourier transform to produce the final WVD spectrum.

The WVD definition in the time domain, formulated in Eq. (6.46), can be implemented to generate the frequency spectrum at any specified time. Substituting $u = t + \tau/2$, the WVD definition can be reformulated as follows (Eq. 6.79):

$$\tilde{S}_{\mathrm{WVD}}(t, f) = 2e^{i2\pi(2f)t} \int_{-\infty}^{\infty} \tilde{z}(u)\tilde{z}^*(2t - u)e^{-i2\pi(2f)u} du, \tag{8.29}$$

where $e^{i2\pi(2f)t}$ is the effect of the time lag t. The implementation of the WVD-based local power spectrum thus consists of the following three main steps.

The first step is to prepare the analytic signal for the calculation of the half-step-length instantaneous autocorrelation. Resampling the analytic signal $z(t) \to z(u)$, with half the original time interval $\Delta u = \Delta t/2$, is achieved by the fast Fourier transform (FFT). The resampled trace $z(u)$ is used for the following two steps, for any time position t.

The second step is to calculate the instantaneous autocorrelation using the multiple orthogonal windows:

(1) Multiplication of the analytic signal $z(u)$ by a window function $w_p(u - t)$,

$$\tilde{z}_p(u;t) = w_p(u - t)z(u). \tag{8.30}$$

(2) Corresponding calculation of the instantaneous autocorrelation,

$$\tilde{\phi}_p(u;t) = \tilde{z}_p(u;t)\tilde{z}_p^*(2t - u;t). \tag{8.31}$$

(3) Averaging of the instantaneous autocorrelations by a weighted sum,

$$\tilde{\phi}(u;t) = \frac{\sum_{p=0}^{P} \lambda_p \tilde{\phi}_p(u;t)}{\sum_{p=0}^{P} \lambda_p}, \tag{8.32}$$

where $P = 4$, the highest order of the multiple orthogonal window functions.

The third step is to generate the WVD-based local power spectrum by Fourier transform:

(1) Performing the Fourier transform,

$$\tilde{\phi}(2f;t) = \mathcal{F}_u\left[\tilde{\phi}(u;t)\right] = \int_{-\infty}^{\infty} \tilde{\phi}(u;t) e^{-i2\pi(2f)u} du. \tag{8.33}$$

(2) Calculating the WVD by

$$\tilde{S}_{\mathrm{WVD}}(t;f) = 2e^{i2\pi(2f)t}\tilde{\phi}(2f;t). \tag{8.34}$$

The second and third steps are repeated for all required time positions.

After completing smoothing filter for the frequency axis, using the multiple orthogonal windows, we can also apply a time-axis smoothing filter $g(t)$ to the WVD spectrum, to further suppress the cross-term interference. In the following example, for the time-axis smoothing filter (Eq. 6.41), the spread parameter is $\sigma_t = 0.012$ s. Meanwhile, for the frequency smoothing filter, the (half-)length of the multiple windows is $T = 0.012$ s.

Figure 8.9 shows five WVDs, corresponding to the five windows with order numbers $p = \{0, 1, \cdots, 4\}$, respectively. When $p = 0$, the WVD clearly indicates a single energy concentration at any given time. Each energy concentration corresponds to a waveform packet in the seismic signal. When $p \geq 1$, each WVD restores some small energy concentrations that were lost in the case of $p = 0$. The energy of these restored concentrations is not necessarily small, otherwise they could be important in seismic characterisation. The summation of the multiple WVDs gives the local power spectrum

Figure 8.10 shows that the local power spectrum produced by the multi-window WVD (Figure 8.10c) is roughly comparable to the spectrum of the smoothed WVD (Figure 8.10b). This similarity is due to the conservation property of the WVD to the frequency support (Section 6.1). However, the Gabor transform method does not have such a finite support property. Therefore, the local power spectrum of the multi-window Gabor transform (Figure 8.8c) is generally different from the local power spectrum of the multi-window WVD (Figure 8.10c).

The local power spectrum in Figure 8.10c shows a weak energy concentration at time 1.55 s. In the smoothed WVD (Figure 8.10b), on the other hand, this weak energy packet was mixed with the strong energy packet at 1.50 s. This weak energy concentration clearly corresponds to the wavelet presented in seismic signal. At the time interval 1.65–1.75 s, two energy concentrations lie next to each other. One has a frequency that gradually decreases over time, and the other has a frequency that gradually increases over time.

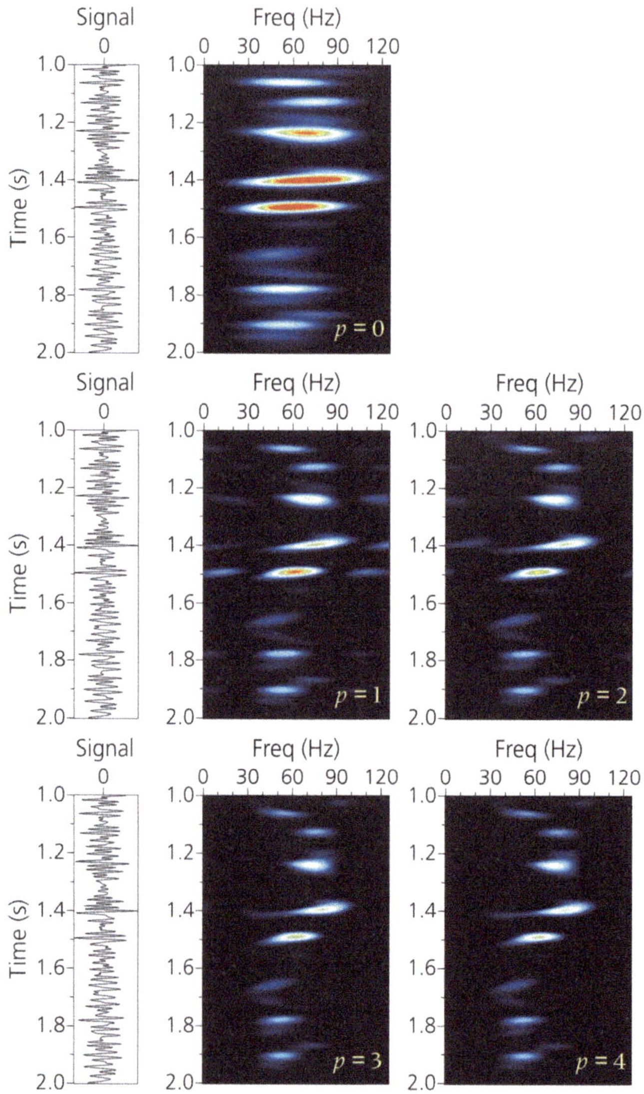

Figure 8.9 The WVD with multiple windows. Each of the multiple orthogonal windows, $p = \{0, 1, \cdots, 4\}$, is used to generate a single WVD. The (half-)length of the multiple windows is $T = 12$ ms.

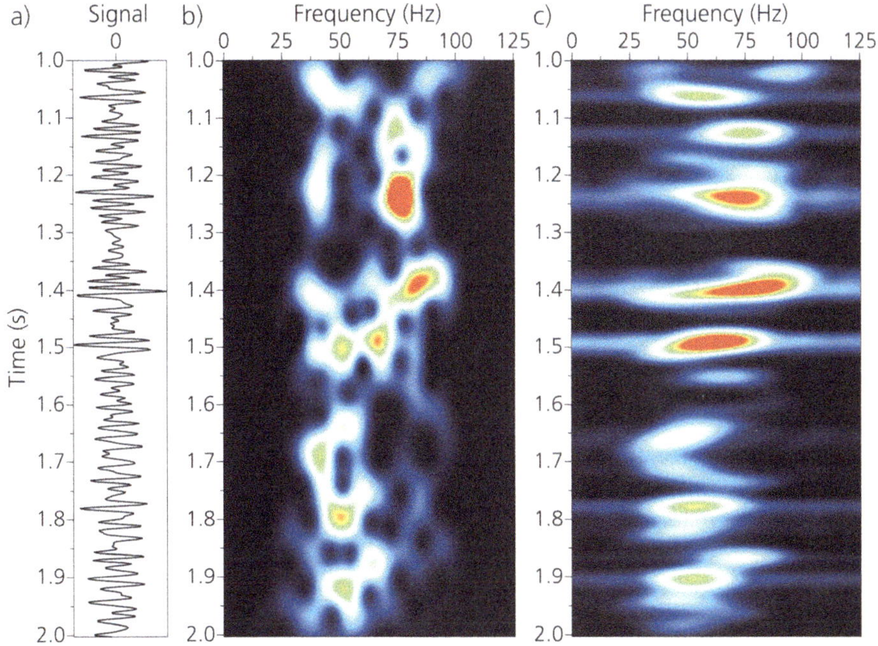

Figure 8.10 Comparison between the smoothed WVD and the multi-window WVD. (a) The field seismic trace. (b) The smoothed WVD, after two-dimensional filtering. (c) Local power spectrum generated by the multi-window WVD.

There are also two energy concentrations in the time interval of 1.75–1.85 s, which were not separated by the smoothed WVD but were mixed at 1.80 s. In the time interval of 1.85–1.95 s, there are three energy concentrations next to each other. The central concentration at 1.9 s has a dominant frequency of 50 Hz, the concentration at the top has a dominant frequency of 70 Hz, and the concentration at the bottom has a dominant frequency of 45 Hz. In this sense, the local power spectrum generated by the multi-window WVD is very close to the matching pursuit-based representation from the previous chapter.

8.7 Prospective of Time–Frequency Analysis without Windowing

There are some methods that attempt to produce the time–frequency spectrum without windowing. One such method is empirical mode decomposition (EMD), which decomposes a signal into several intrinsic mode functions (IMFs) and calculates the instantaneous frequencies for each IMF (Huang *et al.*, 1998). An IMF is a simple oscillating function with varying amplitude and frequency and is symmetric with respect to the local mean zero. The EMD can be viewed as an expansion of the nonstationary signal in terms of IMFs, and these IMFs serve as the basis for the expansion of the signal. However, there are too many different instantaneous frequencies at each time position. This leads to difficulties in interpretation, in addition to the algorithmic ad-hoc nature and lack of mathematical theory about the decomposition.

Basically, an IMF does not necessarily satisfy physical uniqueness because it is sensitive to noise (Meignen & Perrier, 2007; Wu & Huang, 2009). To specifically account for noise in the seismic signal, the decomposition can be reformulated as an inverse problem to minimise the difference between the original signal and the reconstruction from the band-limited IMFs. The constraint for the inverse problem can be formed by the variational of the analytic signal with respect to time and represented in the frequency domain (Dragomiretskiy & Zosso, 2013), so that the centroid of the power spectrum of an IMF can be iteratively re-estimated. Therefore, variational EMD not only improves the robustness of the decomposition to noisy signals, but also overcomes the obvious limitation in terms of sensitivity to sampling.

In the time–frequency representation, the frequency spectrum at each time position can be modelled by a one-dimensional Fourier series. Frequency spectra at different time positions share the same set of Fourier basis functions, but the coefficients are two-dimensional varying with both frequency and time. Then, time–frequency analysis can be reformulated as an inverse problem to estimate these coefficients (Liu *et al.*, 2011; Liu & Fomel, 2013). With suitable regularisations, the window functions need not theoretically be used in this inverse problem. In practice, however, the number of coefficients far exceeds the number of available data samples, and the inverse problem is far underdetermined.

When we use the windowing technique in time–frequency analysis of nonstationary signals, there is a loss of information in the time–frequency spectrum. The local power spectrum generated by WVD with multiple windows successfully presents the details of the signal, thanks to the application of multiple orthogonal windows. Therefore, I am pleased to present the last image on the cover of this book: *Time–Frequency Analysis of Seismic Signals*.

Appendices

A The Gaussian Integrals, the Gamma Function, and the Gaussian Error Functions

This appendix lists various Gaussian integrals commonly used in time–frequency analysis of seismic signals.

The classical Gaussian integral is the definite integral of the Gaussian function e^{-t^2} over the entire real line:

$$\int_{-\infty}^{\infty} e^{-t^2} dt = \sqrt{\pi}. \tag{A.1}$$

The definite integral of an arbitrary Gaussian function $e^{-a(t+b)^2}$ can be derived from the classical Gaussian integral as

$$\int_{-\infty}^{\infty} e^{-a(t+b)^2} dt = \frac{1}{\sqrt{a}} \int_{-\infty}^{\infty} e^{-u^2} du = \sqrt{\frac{\pi}{a}}, \qquad \text{for } a > 0, \tag{A.2}$$

where $u = \sqrt{a}(t+b)$. Because of the symmetry of the Gaussian function, with respect to $t = 0$, the 'half-Gaussian integral' also holds:

$$\int_{0}^{\infty} e^{-at^2} dt = \frac{1}{2}\sqrt{\frac{\pi}{a}}, \qquad \text{for } a > 0. \tag{A.3}$$

For an exponential function with a quadratic exponent, the definite integral can be derived from Eq. (A.2) as

$$\int_{-\infty}^{\infty} e^{-at^2+bt+c} dt = e^{\frac{b^2}{4a}+c} \int_{-\infty}^{\infty} e^{-a\left(t-\frac{b}{2a}\right)^2} dt = \sqrt{\frac{\pi}{a}}\, e^{\frac{b^2}{4a}+c}, \qquad \text{for } a > 0. \tag{A.4}$$

These Gaussian integrals are useful in seismic spectral analysis and processing, in which we often use a window function or a frequency filter in the Gaussian form.

Time–Frequency Analysis of Seismic Signals, First Edition. Yanghua Wang.

Other useful integrals that have a similar form to the Gaussian integral are

$$\int_0^\infty t^{2n} e^{-t^2/a^2} dt = \frac{a^{2n+1}(2n-1)!!}{2^{n+1}} \sqrt{\pi}, \tag{A.5}$$

$$\int_0^\infty t^{2n+1} e^{-t^2/a^2} dt = \frac{n!}{2} a^{2n+2}, \tag{A.6}$$

$$\int_0^\infty t^{2n} e^{-at^2} dt = \frac{(2n-1)!!}{a^n 2^{n+1}} \sqrt{\frac{\pi}{a}}, \tag{A.7}$$

$$\int_0^\infty t^{2n+1} e^{-at^2} dt = \frac{n!}{2a^{n+1}}, \tag{A.8}$$

$$\int_0^\infty t^n e^{-at^2} dt = \frac{1}{2a^{(n+1)/2}} \Gamma\left(\frac{n+1}{2}\right), \tag{A.9}$$

where n is a non-negative integer, $n!$ is the factorial of n, $(2n-1)!!$ denotes the double factorial of an odd number, $(2n-1)!! = (2n-1)\cdot(2n-3)\cdots 3\cdot 1$, and $\Gamma(z)$ is the gamma function defined as

$$\int_0^\infty t^{z-1} e^{-t} dt = \Gamma(z). \tag{A.10}$$

Equation (A.9) shows that the Gaussian integral is related to the gamma function. For example, the classical Gaussian integral of Eq. (A.1) can be represented by the gamma function as follows,

$$\int_{-\infty}^\infty e^{-t^2} dt = 2\int_0^\infty e^{-t^2} dt = \int_0^\infty u^{-1/2} e^{-u} du = \Gamma\left(\frac{1}{2}\right) = \sqrt{\pi}, \tag{A.11}$$

where $u = t^2$. A general form of the gamma function can be derived as follows,

$$\int_0^\infty t^b e^{-at} dt = \frac{1}{a^{b+1}} \int_0^\infty u^b e^{-u} du = \frac{1}{a^{b+1}} \Gamma(b+1), \tag{A.12}$$

where $a > 0$, and $u = at$.

If the Gaussian function is truncated at z, the Gaussian integral is called the Gaussian error function:

$$\operatorname{erf}(z) = \frac{2}{\sqrt{\pi}} \int_0^z e^{-t^2} dt. \tag{A.13}$$

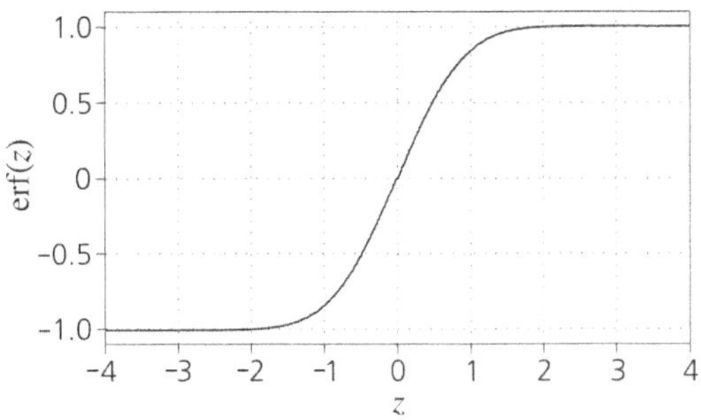

Figure A.1 The Gaussian error function, erf(z). Three characteristic values of the Gaussian error function are $\mathrm{erf}(-\infty) = -1$, $\mathrm{erf}(0) = 0$, and $\mathrm{erf}(\infty) = 1$.

The physical meaning of the Gaussian error function is the ratio of $\int_0^z e^{-t^2} dt$ to the half-Gaussian integral $\int_0^\infty e^{-t^2} dt$, because $\int_0^\infty e^{-t^2} dt = \sqrt{\pi}/2$ (Eq. A.3). Therefore, the characteristic values of the Gaussian error function are

$$\mathrm{erf}(-\infty) = -1, \qquad \mathrm{erf}(0) = 0, \qquad \mathrm{erf}(\infty) = 1. \tag{A.14}$$

As shown in Figure A.1, the Gaussian error function is an odd function:

$$\mathrm{erf}(-z) = -\mathrm{erf}(z). \tag{A.15}$$

This is because the integrand e^{-t^2} is an even function.

A generalised form for the Gaussian error integral between T_1 and T_2 can be derived as follows (Glaisher, 1871a, 1871b)

$$\int_{T_1}^{T_2} e^{-at^2} dt = \int_0^{T_2} e^{-at^2} dt - \int_0^{T_1} e^{-at^2} dt = \frac{1}{\sqrt{a}} \int_0^{\sqrt{a}T_2} e^{-u^2} du - \frac{1}{\sqrt{a}} \int_0^{\sqrt{a}T_1} e^{-u^2} du$$
$$= \frac{1}{2}\sqrt{\frac{\pi}{a}} \left[\mathrm{erf}\left(\sqrt{a}T_2\right) - \mathrm{erf}\left(\sqrt{a}T_1\right) \right]. \tag{A.16}$$

In the Gaussian error function, if the integrand is e^{t^2}, which is an inverse of the truncated Gaussian function e^{-t^2}, the integral is called the imaginary error function:

$$\mathrm{erfi}(z) = \frac{2}{\sqrt{\pi}} \int_0^z e^{t^2} dt. \tag{A.17}$$

Substituting $t = -iu$ into Eq. (A.17), it is easy to prove that the imaginary error function $\mathrm{erfi}(z)$ is related to the Gaussian error function by

$$\mathrm{erfi}(z) = -i\,\mathrm{erf}(iz). \tag{A.18}$$

The Gaussian error function $\mathrm{erf}(z)$ and the imaginary error function $\mathrm{erfi}(z)$ apply to both a real and a complex variable z. They can be evaluated by the Maclaurin series as

$$\mathrm{erf}(z) = \frac{2}{\sqrt{\pi}} \sum_{n=0}^{\infty} \frac{(-1)^n z^{2n+1}}{n!(2n+1)}, \tag{A.19}$$

$$\mathrm{erfi}(z) = \frac{2}{\sqrt{\pi}} \sum_{n=0}^{\infty} \frac{z^{2n+1}}{n!(2n+1)}. \tag{A.20}$$

The Gaussian error function and the imaginary error function are useful in signal analysis, because the time duration of a signal to be analysed is limited. They can be used for example to define the Fresnel integrals:

$$\int_0^z \cos(t^2)\mathrm{d}t = \frac{1-\mathrm{i}}{4}\sqrt{\frac{\pi}{2}}\left[\mathrm{erf}\left(\frac{1+\mathrm{i}}{\sqrt{2}}z\right) + \mathrm{i}\,\mathrm{erf}\left(\frac{1-\mathrm{i}}{\sqrt{2}}z\right)\right], \tag{A.21}$$

$$\int_0^z \sin(t^2)\mathrm{d}t = \frac{1+\mathrm{i}}{4}\sqrt{\frac{\pi}{2}}\left[\mathrm{erf}\left(\frac{1+\mathrm{i}}{\sqrt{2}}z\right) - \mathrm{i}\,\mathrm{erf}\left(\frac{1-\mathrm{i}}{\sqrt{2}}z\right)\right]. \tag{A.22}$$

These Fresnel integrals may also lead to two special integrals, $\int_0^z \mathrm{e}^{-\mathrm{i}t^2}\mathrm{d}t$ and $\int_0^z \mathrm{e}^{\mathrm{i}t^2}\mathrm{d}t$, straightforwardly.

B Fourier Transforms of the Tapered Boxcar Window, the Truncated Gaussian Window, and the Weighted Cosine Window

B.1 The Fourier Transform of the Tapered Boxcar Window

The boxcar window, which is used in the short-time Fourier transform (STFT), can be tapered as

$$w(t) = \frac{1}{r}\begin{cases} \mathrm{e}^{-\frac{(t+T)^2}{2\sigma^2}}, & \text{for } -\infty < t < -T, \\ 1, & \text{for } -T \le t \le T, \\ \mathrm{e}^{-\frac{(t-T)^2}{2\sigma^2}}, & \text{for } T < t < \infty, \end{cases} \tag{B.1}$$

where $T > 0$, and r is a normalisation factor,

$$r = \sqrt{2T + \sqrt{\pi}\sigma}. \tag{B.2}$$

The Fourier transform of the tapered boxcar window is

$$\hat{w}(f)=\frac{1}{r}\int_{-\infty}^{-T}\mathrm{e}^{-\frac{(t+T)^2}{2\sigma^2}-\mathrm{i}2\pi ft}\mathrm{d}t+\frac{1}{r}\int_{-T}^{T}\mathrm{e}^{-\mathrm{i}2\pi ft}\mathrm{d}t+\frac{1}{r}\int_{T}^{\infty}\mathrm{e}^{-\frac{(t-T)^2}{2\sigma^2}-\mathrm{i}2\pi ft}\mathrm{d}t=\frac{1}{r}(I_1+I_2+I_3). \tag{B.3}$$

The first integral I_1 is

$$\begin{aligned}I_1&=\sqrt{2}\sigma\mathrm{e}^{-2\pi^2f^2\sigma^2+\mathrm{i}2\pi fT}\int_{-\infty}^{-T}\mathrm{e}^{-\left(\frac{t+T}{\sqrt{2}\sigma}+\mathrm{i}\sqrt{2}\pi f\sigma\right)^2}\mathrm{d}\left(\frac{t+T}{\sqrt{2}\sigma}+\mathrm{i}\sqrt{2}\pi f\sigma\right)\\&=\sqrt{2}\sigma\mathrm{e}^{-2\pi^2f^2\sigma^2+\mathrm{i}2\pi fT}\int_{-\infty}^{\mathrm{i}\sqrt{2}\pi f\sigma}\mathrm{e}^{-u^2}\mathrm{d}u=\sqrt{2}\sigma\mathrm{e}^{-2\pi^2f^2\sigma^2+\mathrm{i}2\pi fT}\left(\int_{-\infty}^{0}+\int_{0}^{\mathrm{i}\sqrt{2}\pi f\sigma}\right)\mathrm{e}^{-u^2}\mathrm{d}u\\&=\sqrt{\frac{\pi}{2}}\sigma\mathrm{e}^{-2\pi^2f^2\sigma^2+\mathrm{i}2\pi fT}\left[1+\mathrm{erf}(\mathrm{i}\sqrt{2}\pi f\sigma)\right],\end{aligned} \tag{B.4}$$

where $\mathrm{erf}(z)$ is the error function. The second integral I_2 is a scaled sinc function:

$$I_2=\int_{-T}^{T}\mathrm{e}^{-\mathrm{i}2\pi ft}\mathrm{d}t=2T\left(\frac{\sin(2\pi fT)}{2\pi fT}\right). \tag{B.5}$$

The third integral I_3, which is similar to the first integral I_1, is

$$\begin{aligned}I_3&=\sqrt{2}\sigma\mathrm{e}^{-2\pi^2f^2\sigma^2-\mathrm{i}2\pi fT}\int_{T}^{\infty}\mathrm{e}^{-\left(\frac{t-T}{\sqrt{2}\sigma}+\mathrm{i}\sqrt{2}\pi f\sigma\right)^2}\mathrm{d}\left(\frac{t-T}{\sqrt{2}\sigma}+\mathrm{i}\sqrt{2}\pi f\sigma\right)\\&=\sqrt{2}\sigma\mathrm{e}^{-2\pi^2f^2\sigma^2-\mathrm{i}2\pi fT}\int_{\mathrm{i}\sqrt{2}\pi f\sigma}^{\infty}\mathrm{e}^{-u^2}\mathrm{d}u=\sqrt{2}\sigma\mathrm{e}^{-2\pi^2f^2\sigma^2-\mathrm{i}2\pi fT}\left(\int_{0}^{\infty}-\int_{0}^{\mathrm{i}\sqrt{2}\pi f\sigma}\right)\mathrm{e}^{-u^2}\mathrm{d}u\\&=\sqrt{\frac{\pi}{2}}\sigma\mathrm{e}^{-2\pi^2f^2\sigma^2-\mathrm{i}2\pi fT}\left[1-\mathrm{erf}(\mathrm{i}\sqrt{2}\pi f\sigma)\right].\end{aligned} \tag{B.6}$$

Summing these three integrals, the Fourier transform of the tapered boxcar function is

$$\hat{w}(f)=\frac{2T}{r}\left(\frac{\sin(2\pi fT)}{2\pi fT}\right)+\frac{\sqrt{2\pi}\sigma}{r}\mathrm{e}^{-2\pi^2f^2\sigma^2}\left[\cos(2\pi fT)+\mathrm{i}\sin(2\pi fT)\mathrm{erf}(\mathrm{i}\sqrt{2}\pi f\sigma)\right]. \tag{B.7}$$

In this expression, the error function $\mathrm{erf}(\mathrm{i}x)$ with an imaginary variable $\mathrm{i}x$ can be represented as the imaginary error function $\mathrm{erfi}(x)$ with a real variable x, $\mathrm{i}\,\mathrm{erf}(\mathrm{i}x)=-\mathrm{erfi}(x)$, according to Eq. (A.18).

B.2 The Fourier Transform of the Truncated Gaussian Window Function

For the truncated Gaussian window function,

$$w(t) = \frac{1}{\sqrt[4]{\pi\sigma^2}} \begin{cases} e^{-\frac{t^2}{2\sigma^2}}, & \text{for } -T \le t \le T, \\ 0, & \text{otherwise,} \end{cases} \tag{B.8}$$

the Fourier transform is

$$\begin{aligned} \hat{w}(f) &= \frac{1}{\sqrt[4]{\pi\sigma^2}} \int_{-T}^{T} e^{-\frac{t^2}{2\sigma^2} - i2\pi f t} dt = \sqrt[4]{\frac{4\sigma^2}{\pi}} e^{-2\pi^2 f^2 \sigma^2} \int_{-T}^{T} e^{-\frac{(t + i2\pi f \sigma^2)^2}{2\sigma^2}} d\left(\frac{t}{\sqrt{2}\sigma} + i\sqrt{2}\pi f \sigma\right) \\ &= \sqrt[4]{\frac{4\sigma^2}{\pi}} e^{-2\pi^2 f^2 \sigma^2} \left(\int_{0}^{\frac{T}{\sqrt{2}\sigma} + i\sqrt{2}\pi f \sigma} e^{-u^2} du + \int_{-\frac{T}{\sqrt{2}\sigma} + i\sqrt{2}\pi f \sigma}^{0} e^{-u^2} du \right) \\ &= \sqrt[4]{\frac{\pi\sigma^2}{4}} e^{-2\pi^2 f^2 \sigma^2} \left[\operatorname{erf}\left(\frac{T}{\sqrt{2}\sigma} + i\sqrt{2}\pi f \sigma\right) - \operatorname{erf}\left(-\frac{T}{\sqrt{2}\sigma} + i\sqrt{2}\pi f \sigma\right) \right] \\ &= \sqrt[4]{4\pi\sigma^2} e^{-2\pi^2 f^2 \sigma^2} \operatorname{Re}\left[\operatorname{erf}\left(\frac{T}{\sqrt{2}\sigma} + i\sqrt{2}\pi f \sigma\right) \right]. \end{aligned} \tag{B.9}$$

Here, the Gaussian error function erf(z) has a complex variable z, and the derivation exploits the following two properties of the Gaussian error function:

$$\operatorname{erf}(-z) = -\operatorname{erf}(z), \qquad \operatorname{erf}(z^*) = [\operatorname{erf}(z)]^*. \tag{B.10}$$

Because of this conjugate property, the Fourier transform of the truncated Gaussian window function is real.

B.3 The Fourier Transform of the Weighted Cosine Window Function

The weighted cosine window function is defined as

$$w(t) = \begin{cases} \dfrac{1}{\sqrt{\left(c_0^2 + \sum_{k=0}^{K-1} c_k^2\right) T}} \displaystyle\sum_{k=0}^{K-1} c_k \cos\frac{k\pi t}{T}, & \text{for } -T \le t \le T, \\ 0, & \text{otherwise.} \end{cases} \tag{B.11}$$

This continuous window function can be considered as the product of two functions:

$$w(t) = h(t) b(t), \tag{B.12}$$

where

$$h(t) = \frac{1}{\sqrt{\left(c_0^2 + \sum_{k=0}^{K-1} c_k^2\right) T}} \sum_{k=0}^{K-1} c_k \cos\frac{k\pi t}{T}, \tag{B.13}$$

and

$$b(t) = \begin{cases} 1, & -T \le t \le T, \\ 0, & \text{otherwise.} \end{cases} \tag{B.14}$$

Such a product of two functions can be expressed as a convolution in the frequency domain, as

$$\hat{w}(f) = \hat{h}(f) * \hat{b}(f), \tag{B.15}$$

where $\hat{h}(f)$ is the Fourier transform of $h(t)$:

$$\begin{aligned} \hat{h}(f) &= \frac{1}{\sqrt{\left(c_0^2 + \sum_{k=0}^{K-1} c_k^2\right) T}} \sum_{k=0}^{K-1} \int_{-\infty}^{\infty} \frac{c_k}{2}\left(e^{ik\pi t/T} + e^{-ik\pi t/T}\right) e^{-i2\pi ft} dt \\ &= \frac{1}{\sqrt{\left(c_0^2 + \sum_{k=0}^{K-1} c_k^2\right) T}} \sum_{k=0}^{K-1} \frac{c_k}{2}\left[\delta\left(f - \frac{k}{2T}\right) + \delta\left(f + \frac{k}{2T}\right)\right], \end{aligned} \tag{B.16}$$

and $\hat{b}(f)$ is the Fourier transform of $b(t)$:

$$\hat{b}(f) = 2T\left(\frac{\sin(2\pi fT)}{2\pi fT}\right), \tag{B.17}$$

which is a scaled sinc function. Since convolution with an impulse train gives a sum of shifted sinc functions, the Fourier transform $\hat{w}(f)$ of the weighted cosine window function is

$$\hat{w}(f) = \sqrt{\frac{T}{c_0^2 + \sum_{k=0}^{K-1} c_k^2}} \sum_{k=0}^{K-1} c_k \left(\frac{\sin(2\pi fT - k\pi)}{2\pi fT - k\pi} + \frac{\sin(2\pi fT + k\pi)}{2\pi fT + k\pi}\right). \tag{B.18}$$

C The Generalised Seismic Wavelet in the Time Domain

The generalised seismic wavelet is defined as a fractional derivative of a potential function with respect to time, $g^{(u)}(t) = d^u g(t) / dt^u$, where $g(t)$ is the potential function, and u is the order of derivative in the time domain.

We are familiar with the derivatives of the integer orders $g^{(m)}(t)$, where m is an integer. For the fractional derivative $g^{(u)}(t)$, the operator d^u / dt^u can be viewed as a transformation operator, which transforms $g(t)$ to $g^{(u)}(t)$ continuously, according to the continuous varying fractional parameter u, as shown in Figure 3.8.

The uth order derivative of $g(t)$ can be expressed as a Laplace convolution of two causal functions:

$$g^{(u)}(t) = \frac{\mathrm{d}^u g(t)}{\mathrm{d}t^u} = h(t) * g^{(m)}(t), \tag{C.1}$$

where $h(t)$ is a convolution operator, m is an integer, $m-1 \le u < m$, and $g^{(m)}(t)$ is the mth order derivative of $g(t)$. For the generalised seismic wavelet, the potential function $g(t)$ is defined in the following Gaussian form (Wang, 2015):

$$g(t) = -\sqrt{\pi} f_0 \mathrm{e}^{-\pi^2 f_0^2 (t-t_0)^2}, \tag{C.2}$$

where t_0 is the time position of the symmetric centre of the Gaussian function, and f_0 is a reference frequency (in Hz). The Gaussian function $g(t)$ is normalised, $\int_{-\infty}^{\infty} |g(t)| \mathrm{d}t = 1$.

For the Laplace convolution, we set $g^{(m)}(t) = 0$, for $t < 0$. For $t \ge 0$, the integer-order derivatives $g^{(m)}(t)$ can be derived analytically as

$$\begin{aligned}
g^{(m=0)}(t) &= g(t), \\
g^{(m=1)}(t) &= g(t)\left[(-2\pi f_0)(t-t_0)\right], \\
g^{(m=2)}(t) &= g(t)\left[(-2\pi f_0) + (2\pi f_0)^2 (t-t_0)^2\right], \\
g^{(m=3)}(t) &= g(t)\left[2(2\pi f_0)^2 (t-t_0) - (2\pi f_0)^3 (t-t_0)^3\right], \\
g^{(m=4)}(t) &= g(t)\left[2(2\pi f_0)^2 - 5(2\pi f_0)^3 (t-t_0)^2 + (2\pi f_0)^4 (t-t_0)^4\right].
\end{aligned} \tag{C.3}$$

They are in the form of the Gaussian function $g(t)$, multiplied by a polynomial.

The convolution operator $h(t)$ is

$$h(t) = \begin{cases} \dfrac{t^{m-u-1}}{\Gamma(m-u)}, & t > 0, \\ 0, & t \le 0, \end{cases} \tag{C.4}$$

where the gamma function is defined as Eq. (A.10).

Given $h(t)$ and $g^{(m)}(t)$, for $m \ge 1$, the Laplace convolution of Eq. (C.1) can be expressed in an integral form as (Caputo, 1967)

$$g^{(u)}(t) = \frac{1}{\Gamma(m-u)} \int_0^t (t-\xi)^{m-u-1} g^{(m)}(\xi) \mathrm{d}\xi, \qquad m-1 \le u < m. \tag{C.5}$$

The nonzero duration of the integer-order derivative $g^{(m)}(t)$ is relatively short, because of the Gaussian enveloping $g(t)$, as shown in Eq. (C.3). However, the nonzero duration of the operator $h(t)$ can be very long, depending on the power of the factor t^{m-u-1}.

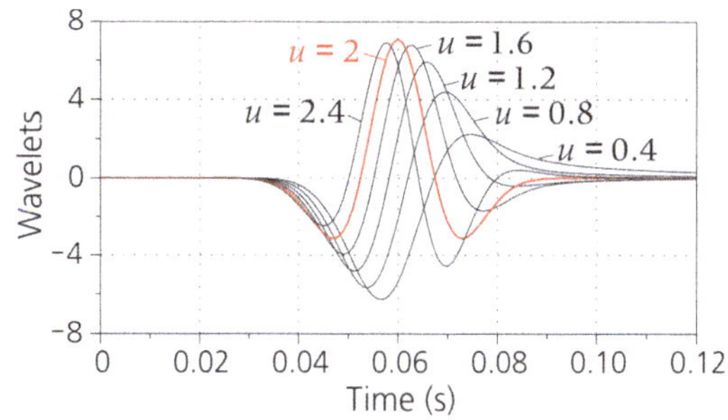

Figure C.1 The generalised seismic wavelets, corresponding to the fractional values $u = \{2.4, 2.0, 1.6, 1.2, 0.8, 0.4\}$. A positive time shift $t_0 = 0.06$ s is set for the calculation of the Laplace convolution.

Note that to generate a generalised seismic wavelet in the time domain, an appropriate positive time shift t_0 in Eq. (C.2) is a necessity. This is because the two causal functions in the Laplace convolution are in the range of $t \geq 0$. In the example shown in Figure C.1, the time shift $t_0 = 0.06$ s is used. The red curve is the Ricker wavelet with the second order derivative ($u = 2$), and the black curves are wavelets defined with different fractional values.

The convolution in the time domain is equivalent to a multiplication between the frequency spectrum of $h(t)$ and the frequency spectrum of $g^{(m)}(t)$. The generalised seismic wavelet in the frequency domain is defined n Section 3.4.

D Implementation of the Fractional Fourier Transform

This appendix summarises the implementation of the forward and inverse fractional Fourier transform (Section 4.3). The forward fractional Fourier transform,

$$\hat{x}_\alpha(u) = \mathcal{F}_\alpha\left[x(t)\right] = \int_{-\infty}^{\infty} x(t) K_\alpha(t, u) \mathrm{d}t, \tag{D.1}$$

can be efficiently calculated using the ordinary Fourier transform:

$$\begin{aligned} \hat{x}_\alpha(u) &= \sqrt{\csc\alpha}\, \mathrm{e}^{-\mathrm{i}(\pi - 2\alpha)/4} \mathrm{e}^{\mathrm{i}\pi u^2 \cot\alpha} \int_{-\infty}^{\infty} \left[x(t) \mathrm{e}^{\mathrm{i}\pi t^2 \cot\alpha} \right] \mathrm{e}^{-\mathrm{i}2\pi(u\csc\alpha)t} \mathrm{d}t \\ &= \sqrt{\csc\alpha}\, \mathrm{e}^{-\mathrm{i}(\pi - 2\alpha)/4} \mathrm{e}^{\mathrm{i}\pi u^2 \cot\alpha} \mathcal{F}_t\left[x(t) \mathrm{e}^{\mathrm{i}\pi t^2 \cot\alpha} \right], \quad (f = u\csc\alpha). \end{aligned} \tag{D.2}$$

The implementation consists of five steps (Erseghe *et al.*, 1999):

(1) Multiplication $x(t)$ by a chirp function: $y(t) = x(t)\mathrm{e}^{\mathrm{i}\pi t^2 \cot\alpha}$.
(2) Performing an ordinary Fourier transform: $\hat{y}(f) = \mathcal{F}_t[y(t)]$.
(3) Converting the frequency of f to u: $\hat{y}(f) \rightarrow \hat{y}_\alpha(u = f\sin\alpha)$.

(4) Multiplying $\hat{y}_\alpha(u)$ by another chirp function: $\hat{z}_\alpha(u) = \hat{y}_\alpha(u)\mathrm{e}^{\mathrm{i}\pi u^2 \cot\alpha}$.
(5) Multiplying $\hat{z}_\alpha(u)$ by a complex factor: $\hat{x}_\alpha(u) = \sqrt{\csc\alpha}\,\mathrm{e}^{-\mathrm{i}(\pi-2\alpha)/4}\hat{z}_\alpha(u)$.

The inverse fractional Fourier transform,

$$x(t) = \mathcal{F}_\alpha^{-1}\left[\hat{x}_\alpha(u)\right] = \int_{-\infty}^{\infty} \hat{x}_\alpha(u)K_\alpha^*(t,u)\mathrm{d}u, \tag{D.3}$$

can also be calculated by the ordinary inverse Fourier transform:

$$\begin{aligned}
x(t) &= \sqrt{\csc\alpha}\,\mathrm{e}^{\mathrm{i}(\pi-2\alpha)/4}\mathrm{e}^{-\mathrm{i}\pi t^2\cot\alpha}\int_{-\infty}^{\infty}\left[\hat{x}_\alpha(u)\mathrm{e}^{-\mathrm{i}\pi u^2\cot\alpha}\right]\mathrm{e}^{\mathrm{i}2\pi u t\csc\alpha}\mathrm{d}u \\
&= \sqrt{\sin\alpha}\,\mathrm{e}^{\mathrm{i}(\pi-2\alpha)/4}\mathrm{e}^{-\mathrm{i}\pi t^2\cot\alpha}\int_{-\infty}^{\infty}\hat{y}_\alpha(u)\mathrm{e}^{\mathrm{i}2\pi(u\csc\alpha)t}\mathrm{d}(u\csc\alpha) \\
&= \sqrt{\sin\alpha}\,\mathrm{e}^{\mathrm{i}(\pi-2\alpha)/4}\mathrm{e}^{-\mathrm{i}\pi t^2\cot\alpha}\int_{-\infty}^{\infty}\hat{y}(f)\mathrm{e}^{\mathrm{i}2\pi f t}\mathrm{d}f \qquad (f = u\csc\alpha) \\
&= \sqrt{\sin\alpha}\,\mathrm{e}^{\mathrm{i}(\pi-2\alpha)/4}\mathrm{e}^{-\mathrm{i}\pi t^2\cot\alpha}\mathcal{F}_f^{-1}\left[\hat{y}(f)\right],
\end{aligned} \tag{D.4}$$

where $\hat{y}_\alpha(u) = \hat{x}_\alpha(u)\mathrm{e}^{-\mathrm{i}\pi u^2\cot\alpha}$. The implementation also consists of five steps, as follows:

(1) Multiplying $\hat{x}_\alpha(u)$ by a chirp function: $\hat{y}_\alpha(u) = \hat{x}_\alpha(u)\mathrm{e}^{-\mathrm{i}\pi u^2\cot\alpha}$.
(2) Converting the frequency of u to f: $\hat{y}_\alpha(u) \to \hat{y}(f = u\csc\alpha)$.
(3) Performing an ordinary inverse Fourier transform: $y(t) = \mathcal{F}_f^{-1}[\hat{y}(f)]$.
(4) Multiplying $y(t)$ by another chirp function: $z(t) = y(t)\mathrm{e}^{-\mathrm{i}\pi t^2\cot\alpha}$.
(5) Multiplying $z(t)$ by a complex factor: $x(t) = \sqrt{\sin\alpha}\,\mathrm{e}^{\mathrm{i}(\pi-2\alpha)/4}z(t)$.

While the operation $\mathcal{F}_t[y(t)]$ is the ordinary forward Fourier transform with respect to time t, the operation $\mathcal{F}_f^{-1}[\hat{y}(f)]$ is the ordinary inverse Fourier transform with respect to frequency f. Both exclude the auxiliary frequency parameter u.

E Marginal Properties and the Analytic Signal in the WVD Definition

E.1 Marginal Properties of the WVD Definition

This appendix proves that the WVD of a real signal $x(t)$ satisfies the two marginal properties given in Eqs. (6.3–6.4).

For a real signal $x(t)$, the WVD is defined as (Wigner, 1932)

$$p_x(t,f) = \int_{-\infty}^{\infty} x\left(t+\frac{\tau}{2}\right)x^*\left(t-\frac{\tau}{2}\right)\mathrm{e}^{-\mathrm{i}2\pi f\tau}\mathrm{d}\tau. \tag{E.1}$$

This is a Fourier transform with respect to the time variable τ. Therefore, the inverse Fourier transform of $p_x(t, f)$ with respect to frequency f can recover the instantaneous autocorrelation:

$$\int_{-\infty}^{\infty} p_x(t, f) e^{i2\pi f\tau} df = x\left(t + \frac{\tau}{2}\right) x^*\left(t - \frac{\tau}{2}\right). \tag{E.2}$$

Transfer the coordinates,

$$t_1 = t + \frac{\tau}{2}, \quad t_2 = t - \frac{\tau}{2}, \tag{E.3}$$

therefore,

$$\tau = t_1 - t_2, \quad t = \frac{t_1 + t_2}{2}. \tag{E.4}$$

We rewrite Eq. (E.2) as

$$\int_{-\infty}^{\infty} p_x\left(\frac{t_1 + t_2}{2}, f\right) e^{i2\pi f(t_1 - t_2)} df = x(t_1) x^*(t_2). \tag{E.5}$$

With the setting $t_1 = t_2 = t$, we immediately obtain the following equation:

$$\int_{-\infty}^{\infty} p_x(t, f) df = x(t) x^*(t) = |x(t)|^2. \tag{E.6}$$

This proves that the WVD satisfies the time marginal property (Eq. 6.3).

Now, we integrate both sides of Eq. (E.1) over time t and obtain a double integral as

$$\int_{-\infty}^{\infty} p_x(t, f) dt = \int_{-\infty}^{\infty}\int_{-\infty}^{\infty} x\left(t + \frac{\tau}{2}\right) x^*\left(t - \frac{\tau}{2}\right) e^{-i2\pi f\tau} d\tau\, dt. \tag{E.7}$$

If we transfer the coordinates according to Eqs. (E.3–E.4), the Jacobi determinant $|\partial(\tau, t)/\partial(t_1, t_2)|$ is equal to one. The double integral in Eq. (E.7) becomes the product of two integrals,

$$\begin{aligned}\int_{-\infty}^{\infty} p_x(t, f) dt &= \int_{-\infty}^{\infty}\int_{-\infty}^{\infty} x(t_1) x^*(t_2) e^{-i2\pi f(t_1 - t_2)} \left|\frac{\partial(\tau, t)}{\partial(t_1, t_2)}\right| dt_1 dt_2 \\ &= \left(\int_{-\infty}^{\infty} x(t_1) e^{-i2\pi f t_1} dt_1\right)\left(\int_{-\infty}^{\infty} x(t_2) e^{-i2\pi f t_2} dt_2\right)^* \\ &= \hat{x}(f)\hat{x}^*(f) = |\hat{x}(f)|^2.\end{aligned} \tag{E.8}$$

This proves that the WVD also satisfies the frequency marginal property (Eq. 6.4).

E.2 The WVD Definition Using an Analytic Signal or a Real Signal

This appendix explains the difference between WVD defined using the analytic signal $z(t)$ and WVD defined using the original real signal $x(t)$. The WVD defined by Eq. (6.7), using the original real signal $x(t)$, does not generally represent the true time–frequency pattern of the signal.

Consider a linear chirp signal $x(t)$ that is expressed as follows (Boashash, 1988)

$$x(t) = b(t)\cos\left[2\pi\left(f_0 t + \frac{\beta}{2}t^2\right)\right], \tag{E.9}$$

where $b(t)$ is the boxcar window function:

$$b(t) = \begin{cases} 1, & \text{for } |t| \le T, \\ 0, & \text{otherwise,} \end{cases} \tag{E.10}$$

and β is the chirp rate, which defines the instantaneous frequency $f_{\text{inst}}(t)$ as

$$f_{\text{inst}}(t) = \frac{1}{2\pi}\frac{\mathrm{d}\theta(t)}{\mathrm{d}t} = f_0 + \beta t. \tag{E.11}$$

The analytic signal $z(t)$ is accordingly,

$$z(t) = b(t)\mathrm{e}^{\mathrm{i}2\pi\left(f_0 t + \frac{\beta}{2}t^2\right)}. \tag{E.12}$$

The WVD of the analytic signal $z(t)$ is

$$\begin{aligned} S_{\text{WVD}}(t,f) &= \int_{-\infty}^{\infty} b\left(t+\frac{\tau}{2}\right)\mathrm{e}^{\mathrm{i}2\pi\left(f_0\left(t+\frac{\tau}{2}\right)+\frac{\beta}{2}\left(t+\frac{\tau}{2}\right)^2\right)} \\ &\quad \times b\left(t-\frac{\tau}{2}\right)\mathrm{e}^{-\mathrm{i}2\pi\left(f_0\left(t-\frac{\tau}{2}\right)+\frac{\beta}{2}\left(t-\frac{\tau}{2}\right)^2\right)}\mathrm{e}^{-\mathrm{i}2\pi f\tau}\mathrm{d}\tau \\ &= \int_{-2(T-|t|)}^{2(T-|t|)} \mathrm{e}^{-\mathrm{i}2\pi[f-f_{\text{inst}}(t)]\tau}\mathrm{d}\tau \\ &= 4(T-|t|)\left(\frac{\sin\left[4\pi[f-f_{\text{inst}}(t)](T-|t|)\right]}{4\pi[f-f_{\text{inst}}(t)](T-|t|)}\right), \end{aligned} \tag{E.13}$$

for $|t| \le T$, $f \le f_{\text{inst}}(t) \pm B$, and B is the bandwidth. The WVD of the analytic signal $z(t)$ has a clear concentration of energy along the line $f = f_{\text{inst}}(t)$, as $f_{\text{inst}}(t)$ is the instantaneous frequency of the linear chirp signal $x(t)$.

To calculate the WVD of the original real signal $x(t)$, the cosine function can be expanded into two complex exponentials as

$$x(t) = \frac{1}{2} b(t) \left(e^{i2\pi\left(f_0 t + \frac{\beta}{2} t^2\right)} + e^{-i2\pi\left(f_0 t + \frac{\beta}{2} t^2\right)} \right). \tag{E.14}$$

Then, the WVD of the real signal $x(t)$ is

$$\begin{aligned} p_x(t,f) = & \frac{1}{4} \int_{-2(T-|t|)}^{2(T-|t|)} e^{i2\pi\left(f_0\left(t+\frac{\tau}{2}\right)+\frac{\beta}{2}\left(t+\frac{\tau}{2}\right)^2\right)} e^{-i2\pi\left(f_0\left(t-\frac{\tau}{2}\right)+\frac{\beta}{2}\left(t-\frac{\tau}{2}\right)^2\right)} e^{-i2\pi f\tau} d\tau \\ & + \frac{1}{4} \int_{-2(T-|t|)}^{2(T-|t|)} e^{-i2\pi\left(f_0\left(t+\frac{\tau}{2}\right)+\frac{\beta}{2}\left(t+\frac{\tau}{2}\right)^2\right)} e^{i2\pi\left(f_0\left(t-\frac{\tau}{2}\right)+\frac{\beta}{2}\left(t-\frac{\tau}{2}\right)^2\right)} e^{-i2\pi f\tau} d\tau \\ & + \frac{1}{4} \int_{-2(T-|t|)}^{2(T-|t|)} e^{i2\pi\left(f_0\left(t+\frac{\tau}{2}\right)+\frac{\beta}{2}\left(t+\frac{\tau}{2}\right)^2\right)} e^{i2\pi\left(f_0\left(t-\frac{\tau}{2}\right)+\frac{\beta}{2}\left(t-\frac{\tau}{2}\right)^2\right)} e^{-i2\pi f\tau} d\tau \\ & + \frac{1}{4} \int_{-2(T-|t|)}^{2(T-|t|)} e^{-i2\pi\left(f_0\left(t+\frac{\tau}{2}\right)+\frac{\beta}{2}\left(t+\frac{\tau}{2}\right)^2\right)} e^{-i2\pi\left(f_0\left(t-\frac{\tau}{2}\right)+\frac{\beta}{2}\left(t-\frac{\tau}{2}\right)^2\right)} e^{-i2\pi f\tau} d\tau \\ = & \frac{1}{4}(I_1 + I_2 + I_3 + I_4), \end{aligned} \tag{E.15}$$

where I_1, I_2, I_3, and I_4 are four integral terms which will be derived in the following.

The first two integral terms, I_1 and I_2, are in similar forms. The first integral term I_1 is given by Eq. (E.13):

$$I_1 = \int_{-2(T-|t|)}^{2(T-|t|)} e^{-i2\pi[f - f_{\text{inst}}(t)]\tau} d\tau = S_{\text{WVD}}(t, f; z), \tag{E.16}$$

and the second integral term I_2 is

$$I_2 = \int_{-2(T-|t|)}^{2(T-|t|)} e^{-i2\pi[f + f_{\text{inst}}(t)]\tau} d\tau = S_{\text{WVD}}(t, -f; z). \tag{E.17}$$

These two integral terms, I_1 and I_2, represent two energy-concentration lines $f = \pm f_{\text{inst}}(t)$, respectively.

The third integral term I_3 is derived as

$$I_3 = e^{i2\pi(2f_0 t+\beta t^2)} \int_{-2(T-|t|)}^{2(T-|t|)} e^{i\frac{\pi\beta\tau^2}{2}} e^{-i2\pi f\tau}\, d\tau$$

$$= e^{i2\pi\left([f_0+f_{\text{inst}}(t)]t-\frac{f^2}{\beta}\right)} \int_{-2(T-|t|)}^{2(T-|t|)} e^{i\frac{\pi\beta}{2}\left(\tau-\frac{2f}{\beta}\right)^2} d\tau = \sqrt{\frac{2}{\pi\beta}}\, e^{i2\pi\left([f_0+f_{\text{inst}}(t)]t-\frac{f^2}{\beta}\right)} \int_{-\sqrt{2\pi\beta}\left(T-|t|+\frac{f}{\beta}\right)}^{\sqrt{2\pi\beta}\left(T-|t|-\frac{f}{\beta}\right)} e^{iu^2} du$$

$$= \frac{1+i}{2\sqrt{\beta}} e^{i2\pi\left([f_0+f_{\text{inst}}(t)]t-\frac{f^2}{\beta}\right)} \left\{ \operatorname{erf}\left[(1-i)\sqrt{\pi\beta}\left(T-|t|-\frac{f}{\beta}\right)\right] + \operatorname{erf}\left[(1-i)\sqrt{\pi\beta}\left(T-|t|+\frac{f}{\beta}\right)\right]\right\}. \tag{E.18}$$

The fourth integral I_4 is derived in a similar way as the first integral I_3, as

$$I_4 = e^{-i2\pi(2f_0 t+\beta t^2)} \int_{-2(T-|t|)}^{2(T-|t|)} e^{-i\frac{\pi\beta\tau^2}{2}} e^{-i2\pi f\tau} d\tau = \sqrt{\frac{2}{\pi\beta}}\, e^{-i2\pi\left([f_0+f_{\text{inst}}(t)]t-\frac{f^2}{\beta}\right)} \int_{-\sqrt{2\pi\beta}\left(T-|t|-\frac{f}{\beta}\right)}^{\sqrt{2\pi\beta}\left(T-|t|+\frac{f}{\beta}\right)} e^{-iu^2} du$$

$$= \frac{1-i}{2\sqrt{\beta}} e^{-i2\pi\left([f_0+f_{\text{inst}}(t)]t-\frac{f^2}{\beta}\right)} \left\{ \operatorname{erf}\left((1+i)\sqrt{\pi\beta}\left(T-|t|+\frac{f}{\beta}\right)\right) + \operatorname{erf}\left((1+i)\sqrt{\pi\beta}\left(T-|t|-\frac{f}{\beta}\right)\right)\right\}. \tag{E.19}$$

For these two derivations, I have derived two special integrals:

$$\int_0^z e^{iu^2} du = \frac{1-i}{\sqrt{2}} \int_0^z e^{\frac{(1+i)^2}{2}u^2} d\left(\frac{1+i}{\sqrt{2}}u\right) = \frac{1-i}{\sqrt{2}} \int_0^{(1+i)z/\sqrt{2}} e^{v^2} dv$$
$$= \frac{1-i}{2}\sqrt{\frac{\pi}{2}} \operatorname{erfi}\left(\frac{1+i}{\sqrt{2}}z\right) = \frac{1+i}{2}\sqrt{\frac{\pi}{2}} \operatorname{erf}\left(\frac{1-i}{\sqrt{2}}z\right), \tag{E.20}$$

$$\int_0^z e^{-iu^2} du = \frac{1-i}{\sqrt{2}} \int_0^z e^{-\frac{(1+i)^2}{2}u^2} d\left(\frac{1+i}{\sqrt{2}}u\right) = \frac{1-i}{\sqrt{2}} \int_0^{(1+i)z/\sqrt{2}} e^{-v^2} dv$$
$$= \frac{1-i}{2}\sqrt{\frac{\pi}{2}} \operatorname{erf}\left(\frac{1+i}{\sqrt{2}}z\right). \tag{E.21}$$

These two integrals lead directly to the Fresnel integrals of Eqs. (A.21–A.22).

Considering the case $\operatorname{erf}(z) \approx 1$, which is valid for $|z| > 2$, following the property of the error function shown in Figure A.1, the integral terms I_3 and I_4 can be approximated as follows,

$$I_3 \approx \frac{1+i}{\sqrt{\beta}} e^{i2\pi\left([f_0+f_{\text{inst}}(t)]t-\frac{f^2}{\beta}\right)}, \qquad I_4 \approx \frac{1-i}{\sqrt{\beta}} e^{-i2\pi\left([f_0+f_{\text{inst}}(t)]t-\frac{f^2}{\beta}\right)}. \tag{E.22}$$

Because of the complex conjugates, $I_3 = I_4^*$, we obtain

$$I_3 + I_4 \approx \frac{2}{\sqrt{\beta}}\left\{\cos\left[2\pi\left([f_0 + f_{\text{inst}}(t)]\,t - \frac{f^2}{\beta}\right)\right] - \sin\left[2\pi\left([f_0 + f_{\text{inst}}(t)]\,t - \frac{f^2}{\beta}\right)\right]\right\}$$

$$= 2\sqrt{\frac{2}{\beta}}\cos\left[2\pi\left([f_0 + f_{\text{inst}}(t)]\,t - \frac{f^2}{\beta}\right) + \frac{\pi}{4}\right]. \tag{E.23}$$

Finally, $p_x(t, f)$, the WVD of the real signal $x(t)$ is

$$p_x(t, f) \approx \frac{1}{4}\left[S_{\text{WVD}}(t, f) + S_{\text{WVD}}(t, -f)\right] + \frac{1}{\sqrt{2\beta}}\cos\left[2\pi\left([f_0 + f_{\text{inst}}(t)]t - \frac{f^2}{\beta}\right) + \frac{\pi}{4}\right], \tag{E.24}$$

for $|t| \le T$, and $|f| \le f_{\text{inst}}(t) + B$. The WVD of the original real signal $x(t)$ has an artefact, in addition to the two WVDs, in positive and negative frequency, of the analytic signal $z(t)$.

F The Prolate Spheroidal Wavefunctions, the Associated and the Ordinary Legendre Polynomials

F.1 Prolate Spheroidal Wavefunctions

The prolate spheroidal wavefunctions $U_{m,n}(c, z)$ satisfy the following differential equation (Niven, 1880; Nicholson, 1925; Van Buren & Boisvert, 2002, 2004):

$$\frac{\mathrm{d}}{\mathrm{d}z}\left((1 - z^2)\frac{\mathrm{d}}{\mathrm{d}z}U_{m,n}(c, z)\right) + \left(\chi_{m,n}(c) - c^2 z^2 - \frac{m^2}{1 - z^2}\right)U_{m,n}(c, z) = 0. \tag{F.1}$$

The prolate spheroidal wavefunctions $U_{m,n}(c, z)$ are a set of orthogonal band-limited functions, and can be written as follows (Beu & Câmpeanu, 1983a, 1983b; Kirby, 2006, 2010)

$$U_{m,n}(c, z) = \begin{cases} \sum\limits_{r=0,2}^{\infty} d_{m,n,r}(c) P_{m,m+r}(z), & \text{if } n - m \text{ is even,} \\ \sum\limits_{r=1,3}^{\infty} d_{m,n,r}(c) P_{m,m+r}(z), & \text{if } n - m \text{ is odd,} \end{cases} \tag{F.2}$$

where $P_{m,m+r}(z)$ are the associated Legendre polynomials, for the case $c = 0$, $d_{m,n,r}(c)$ are the coefficients, and r is an integer. If $n - m$ is even, then r is even, $r = \{0, 2, 4, \cdots\}$, and if $n - m$ is odd, then r is odd, $r = \{1, 3, 5, \cdots\}$.

The associated Legendre polynomials $P_{m,n}(z)$ are the solutions to Eq. (F.1), corresponding to $c = 0$. The eigenvalue $\chi_{m,n}(0) = n(n + 1)$, and Eq. (F.1) becomes the following Legendre differential equation (Falloon *et al.*, 2003):

$$\frac{\mathrm{d}}{\mathrm{d}z}\left((1-z^2)\frac{\mathrm{d}}{\mathrm{d}z}P_{m,n}(z)\right)+\left(n(n+1)-\frac{m^2}{1-z^2}\right)P_{m,n}(z)=0. \tag{F.3}$$

The associated Legendre polynomials $P_{m,n}(z)$ hold orthogonality over the interval $z\in[-1,1]$:

$$\int_{-1}^{1} P_{m,\ell}(z)P_{m,k}(z)\mathrm{d}z=0, \qquad \text{for } \ell\neq k, \tag{F.4}$$

$$\int_{-1}^{1}\left[P_{m,n}(z)\right]^2\mathrm{d}z=\left(\frac{2}{2n+1}\right)\frac{(n+m)!}{(n-m)!}. \tag{F.5}$$

The associated Legendre polynomials $P_{m,n}(z)$, for $0\le m<n$, can be calculated using the following recursive formula:

$$P_{m,\ell+1}(z)=\frac{1+2\ell}{1+\ell-m}zP_{m,\ell}(z)-\frac{\ell+m}{1+\ell-m}P_{m,\ell-1}(z), \tag{F.6}$$

where $\ell=m+r$, for $r\ge 0$. The initial polynomials are given in Appendix F.2.

For the solution of the coefficients $d_{m,n,r}$, which form the prolate spheroidal wavefunctions $U_{m,n}(c,z)$, substituting Eq. (F.2) into Eq. (F.1) yields a recursive relation (Flammer, 1957; Hodge, 1970; Kirby, 2006):

$$A_{m,r}d_{m,n,r+2}+(B_{m,r}-\chi_{m,n})d_{m,n,r}+C_{m,r}d_{m,n,r-2}=0, \tag{F.7}$$

where

$$\begin{aligned}
A_{m,r}&=\frac{(2m+r+2)(2m+r+1)}{(2m+2r+3)(2m+2r+5)}c^2,\\
B_{m,r}&=(m+r)(m+r+1)+\frac{2(m+r)(m+r+1)-2m^2-1}{(2m+2r-1)(2m+2r+3)}c^2,\\
C_{m,r}&=\frac{r(r-1)}{(2m+2r-3)(2m+2r-1)}c^2.
\end{aligned} \tag{F.8}$$

Although these three coefficients in the recursive relation do not depend on n, the n dependence is introduced by the eigenvalue solutions of the following real tridiagonal matrix system:

$$\begin{pmatrix}
b_0 & a_0 & & & & \\
c_1 & b_1 & a_1 & & & \\
 & c_2 & b_2 & a_2 & & \\
 & & \ddots & \ddots & \ddots & \\
 & & & c_k & b_k & a_k \\
 & & & & \ddots & \ddots
\end{pmatrix}
\begin{pmatrix} d_0\\ d_1\\ d_2\\ \vdots\\ d_k\\ \vdots\end{pmatrix}
=\chi\begin{pmatrix} d_0\\ d_1\\ d_2\\ \vdots\\ d_k\\ \vdots\end{pmatrix}, \tag{F.9}$$

where

$$a_k = A_{m,r=2k+s},\quad b_k = B_{m,r=2k+s},\quad c_k = C_{m,r=2k+s},\quad d_k = d_{m,r=2k+s},\tag{F.10}$$

and $s = 0$ or 1, if r is even or odd, respectively. Equation (F.9) can be expressed in a matrix–vector form as

$$\mathbf{Td} = \chi\mathbf{d},\tag{F.11}$$

where $\mathbf{T}$ is the tridiagonal matrix, and $\mathbf{d}$ is the vector of $\{d_k\}$.

The tridiagonal matrix can be transferred to a symmetric tridiagonal matrix, $\mathbf{T} \to \mathbf{DTD}^{-1}$, using a similarity transformation. If matrices $\mathbf{T}$ and $\mathbf{DTD}^{-1}$ are similar, Eq. (F.11) becomes

$$[\mathbf{DTD}^{-1}][\mathbf{Dd}] = \chi[\mathbf{Dd}],\tag{F.12}$$

where $\mathbf{D}$ is a diagonal matrix

$$\mathbf{D} = \mathrm{diag}\{\delta_0, \delta_1, \delta_2, \cdots, \delta_k, \cdots\},\tag{F.13}$$

with

$$\begin{aligned}&\delta_0 = 1,\\ &\delta_k = \delta_{k-1}\sqrt{\frac{a_{k-1}}{c_k}} = \sqrt{\frac{(2m+2s+1)}{(2m+2r+1)(2m+s)!}\frac{(2m+r)!}{r!}},\quad \text{for } k\in[1,N].\end{aligned}\tag{F.14}$$

It is easy to verify that $\mathbf{DTD}^{-1}$ is a symmetric tridiagonal matrix:

$$\mathbf{DTD}^{-1} = \begin{pmatrix} b_0 & e_0 & & & & & \\ e_0 & b_1 & e_1 & & & & \\ & e_1 & b_2 & e_2 & & & \\ & & \ddots & \ddots & \ddots & & \\ & & & e_{k-1} & b_k & e_k & \\ & & & & \ddots & \ddots & \end{pmatrix},\tag{F.15}$$

where $e_k = \sqrt{a_k c_{k+1}}$, for $k\in[0, N-1]$. The symmetric tridiagonal matrix can be solved, using the method QL or QR, for the eigenvalues and the corresponding eigenvectors (Wilkinson, 1968; Hoffman & Parlett, 1978).

The eigenvalue problem is solved twice separately for two cases. First, for the case $s = 0$, corresponding to even integers $p = n - m$, we set $N = K$ and obtain $N+1$ eigenvalues with $n = \{m, m+2, \cdots, m+2N-2, m+2N\}$. Second, for the case $s = 1$, corresponding to odd integers $p = n - m$, we set $N = K - 1$ and obtain N eigenvalues with $n = \{m+1, m+3, \cdots, m+2N-3, m+2N-1\}$. Thus, with the interleaving of the two sets of eigenvalues, we obtain a total of $2N+1$ eigenvalues, sorted ascendingly (Hodge, 1970; Kirby, 2006).

The eigenvector $\mathbf{v}_{m,n}(c)$, corresponding to the eigenvalue $\lambda_{m,n}(c)$, determines the coefficients of a prolate spheroidal wavefunction as

$$\mathbf{d}_{m,n}(c) = \mathbf{D}^{-1}\mathbf{v}_{m,n}(c), \tag{F.16}$$

where $\mathbf{d}_{m,n}(c) = \{d_{m,n,r}(c)\}$. This formula is derived from Eq. (F.12).

Solving Eq. (F.12), we obtain the series of eigenvalues $\chi_{0,n}(c)$, for $m = 0$ and $c = 4\pi$, for example, as ordered as

$$\chi_{0,0}(c) \leq \chi_{0,1}(c) \leq \chi_{0,2}(c) \leq \cdots \leq \chi_{0,2N-1}(c) \leq \chi_{0,2N}(c). \tag{F.17}$$

Figure F.1a shows these eigenvalues, which are normalised and scaled into dB. Figures F.1b–c also show the first 21 coefficient vectors of the prolate spheroidal wavefunctions $\mathbf{d}_{0,n}(c)$, which have nonzero elements within a narrow band along the diagonal direction. Each coefficient vector $\mathbf{d}_{0,n}(c)$ is vertically displayed, and the

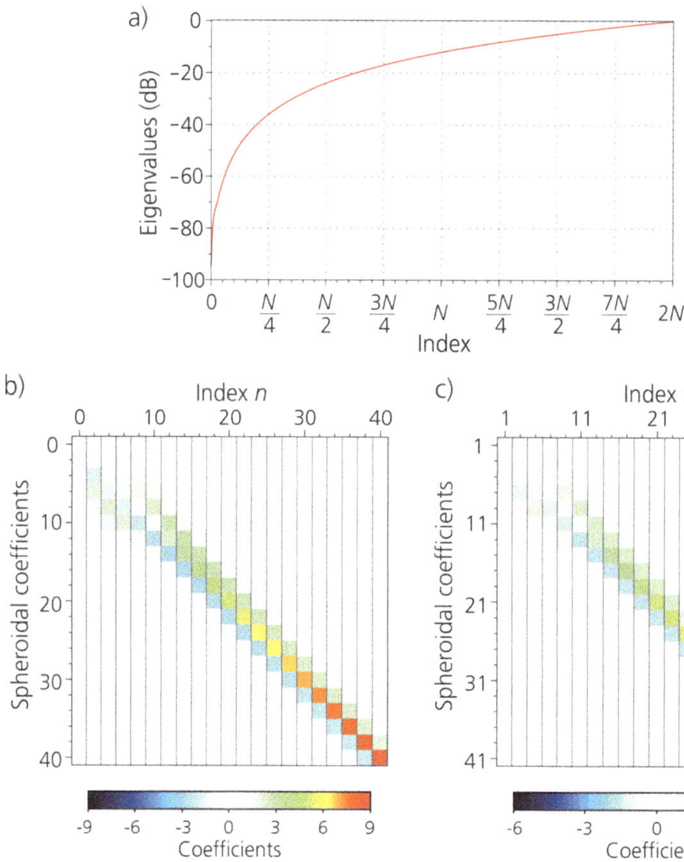

Figure F.1 Solving the coefficients of the prolate spheroidal wavefunctions as an eigenvalue problem. (a) The eigenvalues $\chi_{0,n}(c)$, for $n = \{0, 1, 2, \cdots, 2N-1, 2N\}$, which are sorted into ascending values. The eigenvalues are normalised and scaled in dB. (b) The first 21 coefficient vectors $\mathbf{d}_{0,n}(c)$ corresponding to even indexes $n = \{0, 2, 4, \cdots, 40\}$. (c) The first 21 coefficient vectors $\mathbf{d}_{0,n}(c)$ corresponding to odd indexes $n = \{1, 3, 5, \cdots, 41\}$.

horizontal axis is the index n. Figure F.1b shows the coefficient vectors with even index n, and Figure F.1c shows the coefficient vectors with odd index n. In this example, $K = 400$ and $c = 4\pi$.

F.2 The Associated Legendre Polynomials

The associated Legendre polynomials $P_{m,n}(z)$ are the canonical solutions of the general Legendre equation (Eq. F.3).

The associated Legendre polynomial $P_{m,n}(z)$ can be expressed in the following form:

$$P_{m,n}(z) = \frac{(-1)^m}{2^n n!}(1-z^2)^{m/2}\frac{\mathrm{d}^{m+n}}{\mathrm{d}z^{m+n}}(z^2-1)^n, \tag{F.18}$$

where z is the scaled variable, $-1 \le z \le 1$. The subscript $m \ge 0$ in $P_{m,n}(z)$ indicates the order, and the subscript $n \ge 0$ is the degree of the associated Legendre polynomial. In general, m and n are integers.

The lowest order polynomials are listed as the following.

For $m = 0$:

$$\begin{aligned} P_{0,0}(z) &= 1, \\ P_{0,1}(z) &= z. \end{aligned} \tag{F.19}$$

For $m = 1$:

$$\begin{aligned} P_{1,1}(z) &= -(1-z^2)^{1/2}, \\ P_{1,2}(z) &= -3z(1-z^2)^{1/2}. \end{aligned} \tag{F.20}$$

For $m = 2$:

$$\begin{aligned} P_{2,2}(z) &= 3(1-z^2), \\ P_{2,3}(z) &= 15z(1-z^2). \end{aligned} \tag{F.21}$$

For $m = 3$:

$$\begin{aligned} P_{3,3}(z) &= -15(1-z^2)^{3/2}, \\ P_{3,4}(z) &= -105z(1-z^2)^{3/2}. \end{aligned} \tag{F.22}$$

For $m = 4$:

$$\begin{aligned} P_{4,4}(z) &= 105(1-z^2)^2, \\ P_{4,5}(z) &= 105z(1-z^2)\left[8+(1-9z)(1-z^2)\right]. \end{aligned} \tag{F.23}$$

The high-order polynomials can be derived recursively from the following two formulae:

$$P_{m,\ell+1}(z) = \frac{1+2\ell}{1+\ell-m} z P_{m,\ell}(z) - \frac{\ell+m}{1+\ell-m} P_{m,\ell-1}(z), \tag{F.24}$$

$$P_{m+1,\ell}(z) = \frac{z}{\sqrt{1-z^2}} (\ell-m) P_{m,\ell}(z) - \frac{1}{\sqrt{1-z^2}} (\ell+m) P_{m,\ell-1}(z). \tag{F.25}$$

The term 'associated' Legendre polynomial is a derivation of the ordinary Legendre polynomial:

$$P_{m,n}(z) = (-1)^m (1-z^2)^{m/2} \frac{\mathrm{d}^m}{\mathrm{d}z^m} \hat{P}_n(z), \qquad \text{for } m \geq 0, \tag{F.26}$$

where $\hat{P}_n(z)$ is the ordinary Legendre polynomial from a complete and orthogonal system.

F.3 The Ordinary Legendre Polynomials

The ordinary Legendre polynomials $\hat{P}_n(z)$ are the solutions of the following differential equation:

$$\frac{\mathrm{d}}{\mathrm{d}z}\left((1-z^2) \frac{\mathrm{d}}{\mathrm{d}z} \hat{P}_n(z) \right) + n(n+1)\hat{P}_n(z) = 0. \tag{F.27}$$

This differential equation can be deduced from Eq. (F.3), by setting $m = 0$.

The Legendre polynomial $\hat{P}_n(z)$, for $|z| \leq 1$, is defined as the coefficient of the term t^n in a generating function,

$$\frac{1}{\sqrt{1-2zt+t^2}} = \sum_{n=0}^{\infty} \hat{P}_n(z) t^n, \qquad \text{for } |t| < 1. \tag{F.28}$$

The Legendre polynomial $\hat{P}_n(z)$ has n zeros in the interval $(-1, 1)$, and therefore n is referred to as the degree of the Legendre polynomial.

The Legendre polynomial $\hat{P}_n(z)$ follows a recurrence relationship:

$$(n+1)\hat{P}_{n+1}(z) = (2n+1) z \hat{P}_n(z) - n\hat{P}_{n-1}(z). \tag{F.29}$$

The polynomials with the lowest degrees are $\hat{P}_0(z) = 1$, $\hat{P}_1(z) = z$, which can be derived directly from the Taylor expansion of Eq. (F.28) to $t^{n=1}$.

The recurrence relationship of Eq. (F.29) can be derived from the generating function of Eq. (F.28). Differentiating both sides of Eq. (F.28), with respect to the variable t,

$$\frac{z-t}{\sqrt{1-2zt+t^2}} = (1-2zt+t^2) \sum_{n=0}^{\infty} n\hat{P}_n(z) t^{n-1}, \tag{F.30}$$

and substituting the quotient of the square root with the definition of Eq. (F.28), we get

$$(z-t)\sum_{n=0}^{\infty}\hat{P}_n(z)t^n=(1-2zt+t^2)\sum_{n=0}^{\infty}n\hat{P}_n(z)t^{n-1}. \tag{F.31}$$

Expanding this equation,

$$z\sum_{n=0}^{\infty}\hat{P}_n(z)t^n-\sum_{n=0}^{\infty}\hat{P}_n(z)t^{n+1}=\sum_{n=0}^{\infty}n\hat{P}_n(z)t^{n-1}-2z\sum_{n=0}^{\infty}n\hat{P}_n(z)t^n+\sum_{n=0}^{\infty}n\hat{P}_n(z)t^{n+1}, \tag{F.32}$$

and equating the general coefficients of term t^n on both sides, we have

$$z\hat{P}_n(z)-\hat{P}_{n-1}(z)=(n+1)\hat{P}_{n+1}(z)-2zn\hat{P}_n(z)+(n-1)\hat{P}_{n-1}(z). \tag{F.33}$$

Finally, by simplifying the equation, we derive the recurrence relationship of Eq. (F.29), which is the classical three-term relation for the Legendre polynomials.

Equation (F.29) is a pure recurrence relationship for the Legendre polynomials. Moreover, the Legendre polynomials also follows the differential recurrence relationships:

$$(z^2-1)\frac{\mathrm{d}\hat{P}_n(z)}{\mathrm{d}z}=n\left[z\hat{P}_n(z)-\hat{P}_{n-1}(z)\right]=\frac{n(n+1)}{2n+1}\left[\hat{P}_{n+1}(z)-\hat{P}_{n-1}(z)\right]. \tag{F.34}$$

Differentiating the generating function of Eq. (F.28) with respect to z,

$$t(1-2zt+t^2)^{-3/2}=\sum_{n=0}^{\infty}\frac{\mathrm{d}\hat{P}_n(z)}{\mathrm{d}z}t^n, \tag{F.35}$$

and multiplying both sides by $(z-t)$, we obtain

$$t(z-t)(1-2zt+t^2)^{-3/2}=(z-t)\sum_{n=0}^{\infty}\frac{\mathrm{d}\hat{P}_n(z)}{\mathrm{d}z}t^n. \tag{F.36}$$

If we again differentiate Eq. (F.28) with respect to t,

$$(z-t)(1-2zt+t^2)^{-3/2}=\sum_{n=0}^{\infty}n\hat{P}_n(z)t^{n-1}, \tag{F.37}$$

and substitute it into Eq. (F.36), we get

$$t\sum_{n=0}^{\infty}n\hat{P}_n(z)t^{n-1}=(z-t)\sum_{n=0}^{\infty}\frac{\mathrm{d}\hat{P}_n(z)}{\mathrm{d}z}t^n, \tag{F.38}$$

or

$$\sum_{n=0}^{\infty}n\hat{P}_n(z)t^n=z\sum_{n=0}^{\infty}\frac{\mathrm{d}\hat{P}_n(z)}{\mathrm{d}z}t^n-\sum_{n=0}^{\infty}\frac{\mathrm{d}\hat{P}_n(z)}{\mathrm{d}z}t^{n+1}. \tag{F.39}$$

If we equate the general coefficients of the term t^n on both sides, we obtain

$$n\hat{P}_n(z)=z\frac{\mathrm{d}\hat{P}_n(z)}{\mathrm{d}z}-\frac{\mathrm{d}\hat{P}_{n-1}(z)}{\mathrm{d}z}. \tag{F.40}$$

If we now differentiate both sides of Eq. (F.29) with respect to z, we obtain

$$(n+1)\frac{\mathrm{d}\hat{P}_{n+1}(z)}{\mathrm{d}z} = (2n+1)z\frac{\mathrm{d}\hat{P}_n(z)}{\mathrm{d}z} + (2n+1)\hat{P}_n(z) - n\frac{\mathrm{d}\hat{P}_{n-1}(z)}{\mathrm{d}z}. \tag{F.41}$$

Eliminating $z\mathrm{d}\hat{P}_n(z)/\mathrm{d}z$ from Eq. (F.41), using Eq. (F.40), we get

$$(2n+1)\hat{P}_n(z) = \frac{\mathrm{d}\hat{P}_{n+1}(z)}{\mathrm{d}z} - \frac{\mathrm{d}\hat{P}_{n-1}(z)}{\mathrm{d}z}. \tag{F.42}$$

If we subtract Eq. (F.40) from Eq. (F.42),

$$(n+1)\hat{P}_n(z) = \frac{\mathrm{d}\hat{P}_{n+1}(z)}{\mathrm{d}z} - z\frac{\mathrm{d}\hat{P}_n(z)}{\mathrm{d}z}, \tag{F.43}$$

and replace $(n+1)$ by n, we get an equation as

$$n\hat{P}_{n-1}(z) = \frac{\mathrm{d}\hat{P}_n(z)}{\mathrm{d}z} - z\frac{\mathrm{d}\hat{P}_{n-1}(z)}{\mathrm{d}z}. \tag{F.44}$$

If we multiply Eq. (F.40) by z, we get another equation as

$$nz\hat{P}_n(z) = z^2\frac{\mathrm{d}\hat{P}_n(z)}{\mathrm{d}z} - z\frac{\mathrm{d}\hat{P}_{n-1}(z)}{\mathrm{d}z}. \tag{F.45}$$

Then, we subtract Eq. (F.44) from Eq. (F.45), we get the first differential recurrence relationship shown in Eq. (F.34):

$$(z^2-1)\frac{\mathrm{d}\hat{P}_n(z)}{\mathrm{d}z} = n\left[z\hat{P}_n(z) - \hat{P}_{n-1}(z)\right]. \tag{F.46}$$

Equation (F.29) can be rewritten as

$$(n+1)\left[\hat{P}_{n+1}(z) - z\hat{P}_n(z)\right] = n\left[z\hat{P}_n(z) - \hat{P}_{n-1}(z)\right]. \tag{F.47}$$

The combination of Eqs. (F.46–F.47) leads to

$$(z^2-1)\frac{\mathrm{d}\hat{P}_n(z)}{\mathrm{d}z} = (n+1)\left[\hat{P}_{n+1}(z) - z\hat{P}_n(z)\right]. \tag{F.48}$$

If we multiply Eq. (F.46) by $(n+1)$ and Eq. (F.48) by n, and then add them together, we get

$$(2n+1)(z^2-1)\frac{\mathrm{d}\hat{P}_n(z)}{\mathrm{d}z} = n(n+1)\left[\hat{P}_{n+1}(z) - \hat{P}_{n-1}(z)\right]. \tag{F.49}$$

This is the second differential recurrence relationship, shown in Eq. (F.34).

References

Ackroyd, M.H. (1970). Instantaneous spectra and instantaneous frequency. *Proceedings of the IEEE*, **58**(1), 141.

Ackroyd, M.H. (1971). Short-time spectra and time–frequency energy distributions. *The Journal of the Acoustical Society of America*, **50**, 1229–1231.

Addison, P.S., Walker, J.S. & Guido, R. (2009). Time–frequency analysis of biosignals. *IEEE Engineering in Medicine and Biology Magazine*, **28**(5), 14–29.

Albrecht, H.H. (2001). A family of cosine-sum windows for high-resolution measurements. *Proceedings of IEEE International Conference on Acoustics, Speech and Signal Processing*, 3081–3084.

Allen, J.B. (1977). Short term spectral analysis, synthesis, and modification by discrete Fourier transform. *IEEE Transactions on Acoustics, Speech and Signal Processing*, **25**, 235–238.

Allen, J.B. & Rabiner, L.R. (1977). A unified approach to short-time Fourier analysis and synthesis. *Proceedings of the IEEE*, **65**, 1558–1564.

Almeida, L.B. (1994). The fractional Fourier transform and time–frequency representations. *IEEE Transactions on Signal Processing*, **42**, 3084–3091.

Andrle, M. & Rebollo-Neira, L. (2006). A swapping-based refinement of orthogonal matching pursuit strategies. *Signal Processing*, **86**, 480–495.

Andrle, M., Rebollo-Neira, L. & Sagianos, E. (2004). Backward-optimized orthogonal matching pursuit approach. *IEEE Signal Process Letters*, **11**, 705–708.

Auger, F. & Flandrin, P. (1995). Improving the readability of time–frequency and time–scale representations by the reassignment method. *IEEE Transactions on Signal Processing*, **43**, 1068–1089.

Barnes, A.E. (1992). The calculation of instantaneous frequency and instantaneous bandwidth. *Geophysics*, **57**, 1520–1524.

Bastiaans, M.J. (1980). Gabor's expansion of a signal into Gaussian elementary signals. *Proceedings of the IEEE*, **68**, 538–539.

Baydar, N. & Ball, A. (2001). A comparative study of acoustic and vibration signals in detection of gear failures using Wigner–Ville distribution. *Mechanical Systems and Signal Processing*, **15**, 1091–1107.

Bekara, M. (2013). Taper design for block processing of seismic data. *Geophysics*, **78**(3), A19–A22.

Berkhout, A.J. (1984). *Seismic Resolution: Resolving Power of Acoustical Echo Techniques.* Handbook of Geophysical Exploration, Vol. 12. Geophysical Press, London.

Time–Frequency Analysis of Seismic Signals, First Edition. Yanghua Wang.

Beu, T.A. & Câmpeanu, R.I. (1983a). Prolate radial spheroidal wave functions. *Computer Physics Communications*, **30**, 177–185.

Beu, T.A. & Câmpeanu, R.I. (1983b). Prolate angular spheroidal wave functions. *Computer Physics Communications*, **30**, 187–192.

Blackman, R.B. & Tukey, J.W. (1958). *The Measurement of Power Spectra from the Point of View of Communications Engineering*. Dover Publications Inc.

Boashash, B. (1988). Note on the use of the Wigner distribution for time–frequency signal analysis. *IEEE Transactions on Acoustics, Speech, and Signal Processing*, **36**, 1518–1521.

Bracewell, R.N. (1965). *The Fourier Transform and Its Applications*. McGraw-Hill Inc.

Braga, I.L.S. & Moraes, F.S. (2013). High-resolution gathers by inverse *Q* filtering in the wavelet domain. *Geophysics*, **78**(2), V53–V61.

Capobianco, E. (2003). Independent multiresolution component analysis and matching pursuit. *Computational Statistics & Data Analysis*, **42**, 385–402.

Caputo, M. (1967). Linear models of dissipation whose *Q* is almost frequency independent, II. *Geophysical Journal of the Royal Astronomical Society*, **13**, 529–539.

Castagna, J.P., Sun, S.J. & Siegfried, R.W. (2003). Instantaneous spectral analysis: Detection of low-frequency shadows associated with hydrocarbons. *The Leading Edge*, **22**(2), 120–127.

Chakraborty, A. & Okaya, D. (1995). Frequency–time decomposition of seismic data using wavelet-based methods. *Geophysics*, **60**, 1906–1916.

Champeney, D.C. (1987). *A Handbook of Fourier Theorems*. Cambridge University Press.

Choi, H. & Williams, W.J. (1989). Improved time–frequency representation of multicomponent signals using exponential kernels. *IEEE Transactions on Acoustics, Speech, and Signal Processing*, **37**, 862–871.

Claasen, T.A.C.M. & Mecklenbräuker, W.F.G. (1980a). The Wigner distribution, a tool for time–frequency signal analysis, part I: Continuous-time signals. *Philips Journal of Research*, **35**, 217–250.

Claasen, T.A.C.M. & Meckidnbräuker, W.F.G. (1980b). The Wigner distribution, a tool for time–frequency signal analysis, part II: Discrete time signals. *Philips Journal of Research*, **35**, 276–300.

Claasen, T.A.C.M. & Meckidnbräuker, W.F.G. (1980c). The Wigner distribution, a tool for time–frequency signal analysis, part III: Relations with other time–frequency signal transformations. *Philips Journal of Research*, **35**, 372–389.

Claasen, T.A.C.M. & Meckidnbräuker, W.F.G. (1981). Time–frequency signal analysis by means of the Wigner distribution. *Proceedings of IEEE International Conference on Acoustics, Speech, and Signal Processing*, **3**, 69–72.

Cohen, L. (1966). Generalized phase-space distribution functions. *Journal of Mathematical Physics*, **7**, 781–786.

Cohen, L. (1989). Time–frequency distributions: A review. *Proceedings of the IEEE*, **77**, 941–981.

Daubechies, I. (1992). *Ten Lectures on Wavelets*. Society for Industrial and Applied Mathematics, Philadelphia, PA.

Daubechies, I. & Paul, T. (1988). Time–frequency localization operators, a geometric phase space approach, II: The use of dilations. *Inverse Problems*, **4**, 661–680.

Debnath, L. & Shah, F.A. (2015). The Wigner–Ville distribution and time–frequency signal analysis. In: *Wavelet Transforms and Their Applications*, Birkhäuser, Boston, 287–336.

Djurović, I., Sejdić, E. & Jiang, J. (2008). Frequency based window width optimization for S-transform. *AEU International Journal of Electronics and Communications*, **62**, 245–250.

Dragomiretskiy, K. & Zosso, D. (2013). Variational mode decomposition. *IEEE Transactions on Signal Processing*, **62**, 531–544.

Du, Z.C., Xu, D.P. & Zhang, J.M. (2016). Fractional S-transform, part 2: Application to reservoir prediction and fluid identification. *Applied Geophysics*, **13**, 343–352.

Durka, P.J., Matysiak, A., Montes, E.M. *et al.* (2005). Multichannel matching pursuit and EEG inverse solutions. *Journal of Neuroscience Methods*, **148**, 49–59.

Ebrom, D. (2004). The low-frequency gas shadow on seismic sections. *The Leading Edge*, **23**, 772.

Erseghe, T., Kraniauskas, P. & Carioraro, G. (1999). Unified fractional Fourier transform and sampling theorem. *IEEE Transactions on Signal Processing*, **47**, 3419–3423.

Falloon, P.E., Abbott, P.C. & Wang, J.B. (2003). Theory and computation of spheroidal wavefunctions. *Journal of Physics A: Mathematical and General*, **36**, 5477–5495.

Flammer, C. (1957). *Spheroidal Wave Functions*. Stanford University Press.

Flandrin, P. & Escudié, B. (1984). An interpretation of the pseudo-Wigner–Ville distribution. *Signal Processing*, **6**, 27–36.

Futterman, W.I. (1962). Dispersive body waves. *Journal of Geophysical Research*, **67**, 5279–5291.

Gabor, D. (1946a). Theory of communication, part 1: The analysis of information. *Journal of the Institute of Electrical Engineers*, **93**, 429–441.

Gabor, D. (1946b). Theory of communication, part 2: The analysis of hearing. *Journal of the Institute of Electrical Engineers*, **93**, 442–445.

Gabor, D. (1946c). Theory of communication, part 3: Frequency compression and expansion. *Journal of the Institute of Electrical Engineers*, **93**, 445–457.

Glaisher, J.W.L. (1871a). On a class of definite integrals. *London, Edinburgh, and Dublin Philosophical Magazine and Journal of Science (Series 4)*, **42**(277), 294–302.

Glaisher, J.W.L. (1871b). On a class of definite integrals, part II. *London, Edinburgh, and Dublin Philosophical Magazine and Journal of Science (Series 4)* **42**(282), 421–436.

Goupillaud, P., Grossmann, A. & Morlet, J. (1984). Cycle-octave and related transforms in seismic signal analysis. *Geoexploration*, **23**, 85–102.

Grossmann, A. & Morlet, J. (1984). Decomposition of Hardy functions into square integrable wavelets of constant shape. *SIAM Journal on Mathematical Analysis*, **15**, 723–736.

Hahn, S.L. (1996). *Hilbert Transforms in Signal Processing*. Boston Artech House.

Hall, M.T. (2006). Resolution and uncertainty in spectral decomposition. *First Break*, **24**(12), 43–47.

Hann, J.F. (1903). *Handbook of Climatology, Part I: General Climatology*. Macmillan, New York.

Harris, F.J. (1978). On the use of windows for harmonic analysis with the discrete Fourier transform. *Proceedings of the IEEE*, **66**, 51–83.

Heisenberg, W. (1927). Über den anschaulichen Inhalt der quantentheoretischen Kinematik und Mechanik, (The actual content of quantum theoretical kinematics and mechanics). *Zeitschrift für Physik*, **43**, 172–198.
Helstrom, C.W. (1966). An expansion of a signal in Gaussian elementary signals. *IEEE Transaction Information Theory*, **12**, 81–82.
Hilbert, D. (1912). *Grundzüge Einer Allgemeinen Theorie der Linearen Integralgleichungen*, (*Framework for a General Theory of Linear Integral Equations*). B.G. Teubner, Leipzig & Berlin.
Hlawatsch, F. & Boudreaux-Bartels, G.F. (1992). Linear and quadratic time–frequency signal representations. *IEEE Signal Processing Magazine*, **9**(2), 21–67.
Hodge, D.B. (1970). Eigenvalues and eigenfunctions of the spheroidal wave equation. *Journal of Mathematical Physics*, **11**, 2308–2312.
Hoffman, W. & Parlett, B.N. (1978). A new proof of global convergence for the tridiagonal QL algorithm. *SIAM Journal on Numerical Analysis*, **15**, 929–937.
Holschneider, M. (1995). *Wavelet: An Analysis Tool*. Clarendon Press, Oxford.
Huang, N. E., Shen, Z., Long, S. R. *et al.* (1998). The empirical mode decomposition method and the Hilbert spectrum for nonlinear and nonstationary time series analysis. *Proceedings of the Royal Society of London, Series A*, **454**, 903–995.
Jones, D.L. & Parks, T.W. (1992). A resolution comparison of several time–frequency representations. *IEEE Transactions on Signal Processing*, **40**, 413–420.
Kirby, P. (2006). Calculation of spheroidal wave functions. *Computer Physics Communications*, **175**, 465–472.
Kirby, P. (2010). Calculation of radial prolate spheroidal wave functions of the second kind. *Computer Physics Communications*, **181**, 514–519.
Kolmogorov, A.N. (1950). *Foundations of the Theory of Probability*. Chelsea Publishing Company, New York.
Kolsky, H. (1956). The propagation of stress pulses in viscoelastic solids. *Philosophical Magazine*, **1**, 693–710.
Komorowski, D. & Pietraszek, S. (2016). The use of continuous wavelet transform based on the fast Fourier transform in the analysis of multichannel electrogastrography recordings. *Journal of Medical Systems*, **40**, article no. 10, 1–15.
Korneev, V.A., Goloshubin, G.M., Daley, T.M. *et al.* (2004). Seismic low frequency effects in monitoring fluid-saturated reservoirs. *Geophysics*, **69**, 522–532.
Levin, M. (1964). Instantaneous spectra and ambiguity functions. *IEEE Transactions on Information Theory*, **10**, 95–97.
Li, R., Zhu, X., Zhou, Y. *et al.* (2022). Generalized W transform and its application in gas-bearing reservoir characterization. *IEEE Geoscience and Remote Sensing Letters*, **19**, article no. 3004205, 1–5.
Lilly, J.M. & Olhede, S.C. (2009). High-order properties of analytic wavelets. *IEEE Transactions on Signal Processing*, **57**, 146–160.
Lilly, J.M. & Olhede, S.C. (2010). On the analytic wavelet transform. *IEEE Transactions on Information Theory*, **56**, 4135–4156.
Lilly, J.M. & Olhede, S.C. (2012). Generalized Morse wavelets as a superfamily of analytic wavelets. *IEEE Transactions on Signal Processing*, **60**, 6036–6041.
Lilly, J.M. (2017). Element analysis: A wavelet-based method for analysing time-localized events in noisy time series. *Proceedings of the Royal Society A: Mathematical, Physical and Engineering Sciences*, **473**, article no. 20160776, 1–28.

Liu, G.C., Fomel, S. & Chen, X.H. (2011). Time–frequency analysis of seismic data using local attributes. *Geophysics*, **76**(6), P23–P34.

Liu, J.L. & Marfurt, K.J. (2007). Instantaneous spectral attributes to detect channels. *Geophysics*, **72**, 23–31.

Liu, L.T. & Hsu, H. (2012). Inversion and normalization of time–frequency transform. *Applied Mathematics & Information Sciences*, **6**(1S), 67S–74S.

Liu, Y. & Fomel, S. (2013). Seismic data analysis using local time–frequency decomposition. *Geophysical Prospecting*, **61**, 516–525.

Luo, C.P., Zong, Z.Y. & Xu, L. (2022). Improved W-transform incorporating fast matching pursuit decomposition. *IEEE Geoscience and Remote Sensing Letters*, **19**, article no. 8023905, 1–5.

Mallat, S.G. (2009). *A Wavelet Tour of Signal Processing: The Sparse Way (Third Edition)*. Academic Press.

Mallat, S.G. & Zhang, Z.F. (1993). Matching pursuits with time–frequency dictionaries. *IEEE Transactions on Signal Processing*, **41**, 3397–3415.

Mansinha, L., Stockwell, R.G. & Lowe, R.P. (1997). Pattern analysis with two-dimensional spectral localisation: Applications of two-dimensional S transforms. *Physica A: Statistical Mechanics and its Applications*, **239**, 286–295.

Mark, W.D. (1970). Spectral analysis of the convolution and filtering of non-stationary stochastic processes. *Journal of Sound and Vibration*, **11**, 19–63.

McBride, A.C. & Kerr, F.H. (1987). On Namias's fractional Fourier transform. *IMA Journal of Applied Mathematics*, **39**, 159–175.

Meignen, S. & Perrier, V. (2007). A new formulation for empirical mode decomposition based on constrained optimization. *IEEE Signal Processing Letters*, **14**, 932–935.

Montejo, L.A. & Suarez, L.E. (2013). An improved CWT-based algorithm for the generation of spectrum-compatible records. *International Journal of Advanced Structural Engineering*, **5**, article no. 26, 1–7.

Morlet, J., Arens, G., Fourgeau, E. *et al.* (1982a). Wave propagation and sampling theory, part I: Complex signal and scattering in multilayered media. *Geophysics*, **47**, 203–221.

Morlet, J., Arens, G., Fourgeau, E. *et al.* (1982b). Wave propagation and sampling theory, part II: Sampling theory and complex waves. *Geophysics*, **47**, 222–236.

Namias, V. (1980). The fractional Fourier transform and its application in quantum mechanics. *IMA Journal of Applied Mathematics*, **25**, 241–265.

Nicholson, J.W. (1925). Spheroidal wave-functions. *Proceedings of the Royal Society London A: Mathematical, Physical and Engineering Sciences*, **107**, 43–60.

Niven, C. (1880). On the conduction of heat in ellipsoids of revolution. *Philosophical Transactions of the Royal Society of London*, **171**, 117–151.

Nuttall, A.H. (1981). Some windows with very good sidelobe behavior. *IEEE Transactions on Acoustics, Speech, and Signal Processing*, **29**, 84–91.

Olhede, S.C. & Walden, A.T. (2002). Generalized Morse wavelets. *IEEE Transactions on Signal Processing*, **50**, 2661–2670.

Page, C.H. (1952). Instantaneous power spectra. *Journal of Applied Physics*, **23**, 103–106.

Park, J., Lindberg, C.R. & Thomson, D.J. (1987). Multiple taper spectral analysis of terrestrial free oscillations, part I. *Geophysical Journal of the Royal Astronomical Society*, **91**, 755–794.

Pati, Y.C., Rezaiifar, R. & Krishnaprasad, P.S. (1993). Orthogonal matching pursuit: Recursive function approximation with applications to wavelet decomposition. *Proceedings of 27th Asilomar Conference on Signals, Systems and Computers*, 40–44.

Pinnegar, C.R. & Mansinha, L. (2003). The S-transform with windows of arbitrary and varying shape. *Geophysics*, **68**, 381–385.

Ranjan, R., Jindal, N. & Singh, A.K. (2020). Fractional S-transform and its properties: A comprehensive survey. *Wireless Personal Communications*, **113**, 2519–2541.

Rebollo-Neira, L. & Lowe, D. (2002). Optimized orthogonal matching pursuit approach. *IEEE Signal Process Letters*, **9**, 137–140.

Rényi, A. (1961). On measures of entropy and information. *Proceedings of the Fourth Berkeley Symposium on Mathematical Statistics and Probability, Volume 1: Contributions to the Theory of Statistics*, 547–561.

Ricker, N. (1943). Further developments in the wavelet theory of seismogram structure. *Bulletin of the Seismological Society of America*, **33**, 197–228.

Ricker, N. (1944). Wavelet functions and their polynomials. *Geophysics*, **9**, 314–323.

Ricker, N. (1953). The form and laws of propagation of seismic wavelets. *Geophysics*, **18**, 10–40.

Rihaczek, A.W. (1968). Signal energy distribution in time and frequency. *IEEE Transactions on Information Theory*, **14**, 369–374.

Rioul, O. & Duhamel, P. (1992). Fast algorithms for discrete and continuous wavelet transforms. *IEEE Transactions on Information Theory*, **38**, 569–586.

Rioul, O. & Vetterli, M. (1991). Wavelets and signal processing. *IEEE Signal Processing Magazine*, **8**(4), 14–38.

Schimmel, M. & Gallart, J. (2005). The inverse S-transform in filters with time–frequency localization. *IEEE Transactions on Signal Processing*, **53**, 4417–4422.

Schwarz, H.A. (1888). Über ein Flächen kleinsten Flächeninhalts betreffendes Problem der Variationsrechnung. *Acta Societatis Scientiarum Fennicae*, **15**, 320–361.

Sejdić, E., Djurović, I. & Jiang, J. (2008). A window width optimized S-transform. *EURASIP Journal on Advances in Signal Processing*, article no. 672941, 1–13.

Sejdić, E., Djurović, I. & Jiang, J. (2009). Time–frequency feature representation using energy concentration: An overview of recent advances. *Digital Signal Processing*, **19**, 153–183.

Selesnick, I.W., Baraniuk, R.G. & Kingsbury, N.G. (2005). The dual-tree complex wavelet transform. *IEEE Signal Processing Magazine*, **22**(6), 123–151.

Shin, Y.S. & Jeon, J.J. (1993). Pseudo Wigner–Ville time–frequency distribution and its application to machinery condition monitoring. *Shock and Vibration*, **1**, 65–76.

Simon, C., Ventosa, S., Schimmel, M. *et al.* (2007). The S-transform and its inverses: Side effects of discretizing and filtering. *IEEE Transactions on Signal Processing*, **55**, 4928–4937.

Singh, A. (2015). Fractional S-transform for boehmians. *Journal of Analysis & Number Theory*, **3**, 103–108.

Slepian, D. (1954). Estimation of signal parameters in the presence of noise. *Transactions of the IRE Professional Group on Information Theory*, **3**(3), 68–89.

Slepian, D. (1964). Prolate spheroidal wave functions, Fourier analysis and uncertainty, IV: Extensions to many dimensions; generalized prolate spheroidal functions. *The Bell System Technical Journal*, **43**, 3009–3057.

Slepian, D. (1978). Prolate spheroidal wave functions, Fourier analysis and uncertainty, V: The discrete case. *The Bell System Technical Journal*, **57**, 1371–1430.

Slepian, D. & Pollak, H.O. (1961). Prolate spheroidal wave functions, Fourier analysis, and uncertainty, I. *The Bell System Technical Journal*, **40**, 43–46.

Stanković, L.J. (2001). Measure of some time–frequency distributions concentration. *Signal Processing*, **81**, 621–631.

Stockwell, R.G., Mansinha, L. & Lowe, R.P. (1996). Localization of the complex spectrum: The S transform. *IEEE Transactions on Signal Processing*, **44**, 998–1001.

Stockwell, R.G. (2007). A basis for efficient representation of the S-transform. *Digital Signal Processing*, **17**, 371–393.

Stutt, C.A. (1964). Some results on real-part/imaginary-part and magnitude-phase relations in ambiguity functions. *IEEE Transactions on Information Theory*, **10**, 321–327.

Taner, M.T., Koehler, F. & Sheriff, R.E. (1979). Complex seismic trace analysis. *Geophysics*, **44**, 1041–1063.

Thomson, D.J. (1982). Spectrum estimation and harmonic analysis. *Proceedings of the IEEE*, **70**, 1055–1096.

Torrence, C. & Compo G.P. (1998). A practical guide to wavelet analysis. *Bulletin of the American Meteorological Society*, **79**, 61–78.

Tropp, J.A. & Gilbert, A.C. (2007). Signal recovery from random measurements via orthogonal matching pursuit. *IEEE Transactions on Information Theory*, **53**, 4655–4666.

Tukey, J.W. & Hamming, R.W. (1949). Measuring noise color. *The Collected Works of John W. Tukey: Time Series 1949-1964, Volume I*. Bell Telephone Laboratories, Inc.

Van Buren, A.L. & Boisvert, J.E. (2002). Accurate calculation of prolate spheroidal radial functions of the first kind and their first derivatives. *Quarterly of Applied Mathematics*, **60**, 589–599.

Van Buren, A.L. & Boisvert, J.E. (2004). Improved calculation of prolate spheroidal radial functions of the second kind and their first derivatives. *Quarterly of Applied Mathematics*, **62**, 493–507.

Ventosa, S., Simon, C., Schimmel, M. *et al.* (2008). The *S*-transform from a wavelet point of view. *IEEE Transactions on Signal Processing*, **56**, 2771–2780.

Vetterli, M. & Herley, C. (1992). Wavelets and filter banks: Theory and design. *IEEE Transactions on Signal Processing*, **40**, 2207–2232.

Ville, J. (1948). Théorie et Applications de la notion de signal analytique (Theory and applications of the notion of the analytic signal). *Cables et Transmissions*, **2**, 61–74.

Wang, Y.H. (1998). Continuous unwrapped phase. *Journal of Seismic Exploration*, **7**, 109–116.

Wang, Y.H. (2000). A robust seismic phase unwrapping method. *Journal of Seismic Exploration*, **9**, 93–104.

Wang, Y.H. (2002). A stable and efficient approach of inverse *Q* filtering. *Geophysics*, **67**, 657–663.

Wang, Y.H. (2006). Inverse *Q* filter for seismic resolution enhancement. *Geophysics*, **71**(3), V51–V60.

Wang, Y.H. (2007). Seismic time–frequency spectral decomposition by matching pursuit. *Geophysics*, **72**(1), V13–V20.

Wang, Y.H. (2008). *Seismic Inverse Q Filtering*. Blackwell Publishing, Oxford.

Wang, Y.H. (2010). Multichannel matching pursuit for seismic trace decomposition. *Geophysics*, **75**(4), V61–V66.

Wang, Y.H. (2015). Generalized seismic wavelets. *Geophysical Journal International*, **203**, 1172–1178.

Wang, Y.H. (2021). The W transform. *Geophysics*, **86**(1), V31–V39.

Wang, Y.H. & Guo, J. (2004). Modified Kolsky model for seismic attenuation and dispersion. *Journal of Geophysics and Engineering*, **1**, 187–196.

Wang, Y.H., Liu, X., Gao, F. *et al.* (2020). Robust vector median filtering with a structure-adaptive implementation. *Geophysics*, **85**(5), V407–V414.

Wigner, E.P. (1932). On the quantum correction for thermodynamic equilibrium. *Physical Review*, **40**, 749–759.

Wilkinson, J.H. (1968). Global convergence of tridiagonal QR algorithm with origin shifts. *Linear Algebra and Its Applications*, **1**, 409–420.

Wilson, R. (1987). Finite prolate spheroidal sequences and their applications I: Generation and properties. *IEEE Transactions on Pattern Analysis and Machine Intelligence*, **9**, 787–795.

Wilson, R. & Spann, M. (1988). Finite prolate spheroidal sequences and their applications, II: Image feature description and segmentation. *IEEE Transactions on Pattern Analysis and Machine Intelligence*, **10**, 193–203.

Wu, Z. & Huang N.E. (2009). Ensemble empirical mode decomposition: A noise-assisted data analysis method. *Advances in Adaptive Data Analysis*, **1**, 1–41.

Xu, D.P. & Guo, K. (2012). Fractional S transform, part 1: Theory. *Applied Geophysics*, **9**, 73–79.

Zaghloul, M.R. & Ali, A.N. (2012). Algorithm 916: Computing the faddeyeva and voigt functions. *ACM Transactions on Mathematical Software*, **38**(2), article no. 15, 1–22.

Zhivomirov, H. (2019). On the development of STFT-analysis and ISTFT-synthesis routines and their practical implementation. *TEM Journal*, **8**, 56–64.

Zielinski, T.P. (2001). Joint time–frequency resolution of signal analysis using Gabor transform. *IEEE Transactions on Instrumentation and Measurement*, **50**, 1436–1444.

Author Index

Ackroyd, M.H. 10, 13
Addison, P.S. 51, 63
Albrecht, H.H. 28, 30
Ali, A.N. 25
Allen, J.B. 21, 40
Almeida, L.B. 80
Andrle, M. 156
Auger, F. 134

Ball, A. 134
Barnes, A.E. 14
Bastiaans, M.J. 47
Baydar, N. 134
Bekara, M. 186
Berkhout, A.J. 34
Beu, T.A. 209
Blackman, R.B. 28
Boashash, B. 206
Boisvert, J.E. 178, 209
Boudreaux-Bartels, G.F. 119
Bracewell, R.N. 7
Braga, I.L.S. 69

Câmpeanu, R.I. 209
Capobianco, E. 152
Caputo, M. 202
Castagna, J.P. 109, 174
Chakraborty, A. 174
Champeney, D.C. 10
Choi, H. 134
Claasen, T.A.C.M. 124–126, 129, 134
Cohen, L. 119, 124, 132, 134
Compo, G.P. 51

Daubechies, I. 52, 54
Debnath, L. 119
Djurović, I. 78
Dragomiretskiy, K. 196
Du, Z.C. 83
Duhamel, P. 67
Durka, P.J. 166

Ebrom, D. 109
Erseghe, T. 203
Escudié, B. 130

Falloon, P.E. 209
Flammer, C. 178, 210
Flandrin, P. 130, 134
Fomel, S. 196
Futterman, W.I. 5–6

Gabor, D. 17, 23, 33, 47, 153
Gallart, J. 88
Gilbert, A.C. 156
Glaisher, J.W.L. 197
Goupillaud, P. 49
Grossmann, A. 49
Guo, J. 5
Guo, K. 83

Hahn, S.L. 11
Hall, M.T. 21, 36
Hamming, R.W. 28
Hann, J.F. 27
Harris, F.J. 26, 28, 30
Heisenberg, W. 36
Helstrom, C.W. 47, 51
Herley, C. 67
Hilbert, D. 11
Hlawatsch, F. 119
Hodge, D.B. 210–211
Hoffman, W. 211
Holschneider, M. 65
Hsu, H. 65
Huang, N.E. 196

Jeon, J.J. 128

Time–Frequency Analysis of Seismic Signals, First Edition. Yanghua Wang.

Jones, D.L. 134

Kerr, F.H. 80
Kirby, P. 209–211
Kolmogorov, A.N. 119
Kolsky, H. 6
Komorowski, D. 51
Korneev, V.A. 174

Levin, M. 136
Li, R. 99
Lilly, J.M. 55
Liu, G.C. 196
Liu, J.L. 152
Liu, L.T. 65
Liu, Y. 196
Lowe, D. 156
Luo, C.P. 157

Mallat, S.G. 50, 52, 152–153
Mansinha, L. 74
Marfurt, K.J. 152
Mark, W.D. 120
McBride, A.C. 80
Mecklenbräuker, W.F.G. 129, 134
Meignen, S. 196
Montejo, L.A. 51
Moraes, F.S. 69
Morlet, J. 49, 157

Namias, V. 80
Nicholson, J.W. 178–179, 209
Niven, C. 178, 209
Nuttall, A.H. 26, 32

Okaya, D. 174
Olhede, S.C. 49, 54–55

Page, C.H. 119
Park, J. 176
Parks, T.W. 134
Parlett, B.N. 211
Pati, Y.C. 156
Paul, T. 54
Perrier, V. 196
Pietraszek, S. 51
Pinnegar, C.R. 74
Pollak, H.O. 178

Rabiner, L.R. 40
Ranjan, R. 83
Rebollo-Neira, L. 156
Rényi, A. 105
Ricker, N. 4, 58–59
Rihaczek, A.W. 136
Rioul, O. 62, 67

Schimmel, M. 88
Schwarz, H.A. 35
Sejdić, E. 78
Selesnick, I.W. 49
Shah, F.A. 119
Shin, Y.S. 128
Simon, C. 74
Singh, A. 83
Slepian, D. 178
Spann, M. 182
Stanković, L.J. 105
Stockwell, R.G. 74, 76
Stutt, C.A. 136
Suarez, L.E. 51

Taner, M.T. 12
Thomson, D.J. 176
Torrence, C. 51
Tropp, J.A. 156
Tukey, J.W. 28

Van Buren, A.L. 178, 209
Ventosa, S. 73
Vetterli, M. 62, 67
Ville, J. 10, 13, 117–118, 121

Walden, A.T. 49, 54
Wang, Y.H. 5–6, 13, 42, 45, 58, 64, 95, 109, 152, 155, 159, 161, 163, 168, 202
Wigner, E.P. 117, 120, 204
Wilkinson, J.H. 211
Williams, W.J. 134
Wilson, R. 182
Wu, Z. 196

Xu, D.P. 83

Zaghloul, M.R. 25
Zhang, Z.F. 152–153
Zhivomirov, H. 41
Zielinski, T.P. 42
Zosso, D. 196

Subject Index

admissible constant 65
admissible wavelet 67
ambiguity function 134–139, 147–150
analytic signal 1, 10–12, 14, 16, 120–126, 141–142, 144, 147–149, 152, 155–156, 191–192, 196, 206, 208, 211
associated Legendre polynomial 211–212, 215–216
autocorrelation 123, 126, 133, 135–137
auxiliary frequency 80, 82–85, 89, 206

bandwidth retention factor 176–178, 180, 186, 189, 191
Blackman window 28–30, 32–33, 39–40, 200, 202–203
Blackman–Harris window 29–30, 32–33
boxcar window 20–24, 31–32, 37–38, 47, 120, 200, 208

Cauchy principal value 11
Cauchy–Schwarz inequality 35
chirp function 90, 203–204
complex energy density 136–137
continuous wavelet transform 17, 47, 49–54, 58, 62, 65, 67–69, 72–74, 95, 106, 117, 152–153, 170–171, 175, 188
correlative interpretation 117, 120
cross-correlation 8–9, 50–51, 72, 126, 152, 172–173, 186–187
cross-term interference 126–128, 130–131, 134, 138, 141–142, 145, 147–148, 151, 174–175, 191, 193

delay time 135, 139, 148–149
differential singularity 100–101
dissipation coefficient 5
doppler frequency 135, 139, 149

elementary signal 47
energetic interpretation 117, 120
energy conservation 10, 50, 81–84, 87, 89–90, 109, 119
energy density distribution 17–18, 118–119, 121, 125, 140, 151, 156, 159, 162, 191

fractional derivative 59–61, 203
fractional Fourier transform 79–87, 89–90, 92, 205–206
fractional S transform 83–85, 87, 89
Fredholm integral 178, 180
Fresnel integral 200, 210
Frobenius norm 168–169

Gabor limit 33–34, 36, 99, 120
Gabor transform 17, 19, 23–24, 36–44, 46–47, 49, 73–74, 77, 96, 99, 102, 106, 117, 120, 151–153, 160, 175, 187, 189–190, 193
Gabor wavelet 47–49, 52, 127, 130–131, 137–139, 153–154, 157, 161, 175, 187
gamma function 15, 197–198, 204
Gaussian error function 22–23, 25, 197–200, 202
Gaussian filter 16
Gaussian integral 15, 197–199
Gaussian spread parameter 15, 40, 52–53, 74, 96, 100, 102, 107, 131
Gaussian window 24–25, 27, 32, 74–76, 78, 83, 86, 88–89, 95–96, 99–101, 106–108, 157, 194
generalised ambiguity function 139–140, 148
generalised autocorrelation 133
generalised S transform 77–79, 84–85, 87–89, 91, 107
generalised seismic wavelet 58–62, 64, 203–205

Time–Frequency Analysis of Seismic Signals, First Edition. Yanghua Wang.

generalised W transform 99–108
generalised WVD 134, 138–140, 147

Hamming window 28, 30–31, 33, 37, 40
Hanning window 27, 30–31, 33, 38
Hilbert transform 10–12, 15, 60

imaginary error function 22–23, 199–201
instantaneous amplitude 12, 14, 16
instantaneous autocorrelation 120–121, 125–126, 135, 141–142, 144, 191–192, 207
instantaneous energy 118, 124, 190
instantaneous frequency 12–16, 36, 120, 122, 124, 155, 157, 208–209
instantaneous phase 12–13, 155, 172
intrinsic mode function 195
inverse fractional Fourier transform 80, 89, 92, 206
inverse Gabor transform 40–43
inverse W transform 108
inverse wavelet transform 65–67, 69
invertibility condition 41, 92

Kolmogorov axiom 119

Laplace convolution 204–205
Legendre differential equation 211
Legendre polynomial 211–212, 215
local power spectrum 175, 190, 192–193, 195–196
low-frequency shadow 93, 109, 111–112, 115, 173–174

marginal property 119,123–124, 205
matching pursuit 151–161, 163–166, 168–172, 174
modified Morse wavelet 55–58
Morlet wavelet 51–53, 58, 63, 68–69, 157–159, 162
Morse wavelet 54–58, 63–64
multichannel matching pursuit 163–166, 168, 171–174
multiple windows 17, 175, 176, 178–181, 183–185, 187, 189, 191, 193–194

nonstationary convolution 5–7, 106, 172
nonstationary signal 1, 5–7, 10, 12, 16, 47, 53, 88, 117
nonstationary wavelet 5–7

orthogonal windows 17, 176, 188–194, 196
orthogonality 156, 212

Parseval's theorem 10, 24, 75, 81, 84, 121, 177
potential function 59, 203–204
prolate spheroidal wavefunction 178–180, 182–184, 211–212, 214–215
pseudo-frequency 62–63, 68, 73, 152, 171
pseudo-WVD 129–130

quadratic representation 117, 119
quasi-probability 119

Rényi entropy 105–106
Ricker wavelet 4, 14–15, 51, 58–59, 97–98, 205
Rihaczek distribution 136

S transform 17, 73–80, 83–84, 86–89, 95, 97–99, 106–107, 109–111, 113–114, 117, 152
short-time Fourier transform 17, 19, 47, 175, 200
sigma filter 159, 161, 163
signum function 11
similarity transformation 213
singularity 69, 76, 87, 95, 107–108
smoothed WVD 126, 130–132, 134, 140, 145–147, 150, 193, 195
smoothing filter 128–132, 134, 140, 145–146, 191, 193
spectral leakage 21, 23–24, 26–27, 31–32, 177, 182, 189
STFT 17, 19–20, 23, 47, 125, 200
structure adaptive 168–170
superposition 1–3, 117

tapered boxcar window 21–23, 31–32, 37–38, 200–201
truncated Gaussian window 25–26, 32, 37, 182, 200, 202

uncertainty principle 36
unitarity 119

W transform 17, 95–101, 103–107, 109, 111–115, 117, 152, 157
Wigner–Ville distribution 17, 117, 151
WVD 17, 117, 119–152, 156, 158, 162, 174–175, 191–196, 206–209, 211